Lean Office and Service Simplified

The Definitive How-to Guide

Endorsements

"Drew is still one of the few practitioners who really understand the concept of the Lean Enterprise as opposed to Lean Manufacturing. The examples in this book are real business examples as opposed to factory floor examples that have just been reworded."—Bill Beer, President, Wenger Corporation

"Office and Services are quickly learning that a Lean approach to everything they do is THE way to remain competitive and deliver value for the customer. Drew Locher has delivered a plain language guide to transforming Office and Service industries into effective, efficient organizations. I recommend it strongly."—Mike Robinson, Manager, Global Training & Organizational Development, Corning Cable Systems

"A true Lean Enterprise must go beyond the shop floor to garner the true benefits of Lean. This book is an excellent summary of the areas of opportunity and the tools that can be deployed in an office or services setting."—Ray Keefe, VP, Manufacturing, Emerson

"Locher addresses the real waste in an organization—the timely flow of information across the value stream. He provides a simple guide to help leaders drive the Lean transformation of themselves, their people and processes."—Glenn Marshall, Benchmarking & Sustainability Champion, Northrop Grumman Shipbuilding

"*Lean Office and Service Simplified* is an excellent primer for those getting started or those making the transition with office or service Lean. It is a succinct survey of Lean applications in administrative, technical, professional and service settings. It should serve as the basis from which new or transitioning practitioners can find their way through a Lean implementation."—David Mann, author of *Creating a Lean Culture: Tools to Sustain Lean Conversions*

"Drew Locher's new book, *Lean Office and Service Simplified,* more than lives up to its title. It is accessible and practical. It is easy to understand yet thoroughly Lean-focused. And it is filled with real-life experience. Mr. Locher provides clear and straightforward methods for applying Lean thinking to administrative and support processes that apply across almost every industry. The explanations of these methods are backed up with examples from specific process like Purchasing or Sales & Marketing, and the book also addresses the leadership and human aspects of Lean transformation in an office setting. I highly recommend this book."—Brian Maskell, President, BMA Inc.

"Whether you're developing your Lean strategy or taking specific techniques to the next level, Locher's book delivers! Every essential tool in the Lean toolkit is explored with enough simplicity for a beginner to understand and enough depth for an experienced Lean Thinker to draw from. The Lean Leadership section alone makes it a compelling read."—Allan R. Coletta, Sr. Director Engineering and Facilities, Siemens Healthcare Diagnostics, Inc.

Lean Office and Service Simplified

The Definitive How-to Guide

Drew Locher

CRC Press is an imprint of the
Taylor & Francis Group, an **informa** business
A PRODUCTIVITY PRESS BOOK

Productivity Press
Taylor & Francis Group
270 Madison Avenue
New York, NY 10016

Printed in the United States of America on acid-free paper
10 9 8 7 6 5 4 3

International Standard Book Number: 978-1-4398-2031-5 (Hardback)

Visit the Taylor & Francis Web site at
http://www.taylorandfrancis.com

and the Productivity Press Web site at
http://www.productivitypress.com

To my wife Eileen. This one was long and difficult. Thank you for your support and patience throughout. To my parents Adele and Walt. I cannot begin to tell you how much I appreciate all that you have done throughout my life. I hope that I have made you proud.

Contents

Endorsements .... **ii**

Introduction .... **xi**

1 Organizing by Value Stream **1**

Cross-Functional Teams Based on Value Stream 3

Defining Roles by Value Stream within a Department 4

Organizing Activities for Individuals by Value Stream 6

Summary 7

2 Creating Standard Work for Office and Service **9**

Standard Work—What It Is and Its Purpose 10

Elements of Standard Work 12

The "What" 12

Key Points—The "How" and "Why" 12

Time and Timing 13

Visually Displaying Standard Work 14

Benefits 16

Steps to Creating Standard Work 17

Summary 19

3 Creating Flow in Office and Services **21**

Approaches to Flow in Office and Services 22

Combining Activities 23

Continuous Flow Processing with Multiple Roles 25

Concurrent Processing 28

Designing Flow Systems for Office and Services 29

Identify Activities 30

Determine the Demand Rate 30

Determine Resource Requirements 32

Identify Roles and Responsibilities, including Standard Work 34

Determine Training and Cross-Training Needs 36
Develop Visual Management Techniques 37
Typical Results 38
Summary 39

4 Creating Level Pull in the Office 41
Forms of Pull Systems 42
Visibility of Queues 45
Establishing Limits on Queues 46
Establishing Decision Rules for the Queue 48
Using Visual Signals That Are Worker Managed 50
Leveling the System 52
Steps to Implement Pull Systems 54
Step 1: Identify the Locations Where Queues Are Expected to Form 55
Step 2: Identify Means to Provide Visibility 55
Step 3: Establish Limits for the Queue 56
Step 4: Define Rules for the Queue 57
Step 5: Train People in the Pull System 57
Step 6: Monitor the System for Effectiveness 57
Benefits of Office and Service Pull Systems 58
Summary 60

5 Establishing Visual Management in Office and Services 61
Background 61
Approaches to Visual Management 63
Elements of Visual Management 64
What Is the Purpose or Function of the Area? 65
What Activities Are Performed in the Area? 65
How Do People Know What To Do? 66
How Do They Know How To Do It? 68
How Do They Know How They Are Doing? 69
What Is Done If Performance Expectations Are Not Being Met? 71
Including Continuous Improvement in Visual Management 72
Summary 73

6 Lean Tools for Office and Services 75
Workplace Organization or 5S 76
Mistake Proofing 84

Terms and Definitions 86
Mistake Proofing Devices 86
Mistake Proofing Devices and Examples 88
Guide/Reference/Interference Device 88
Template/Checklist Devices 88
Light Contact Electrical Devices 88
Counter Devices 89
Odd-Part-Out Devices 89
Sequence Restriction Devices 90
Standardize and Solve Devices 90
Critical Condition Indicator Devices 91
Delivery Detection Devices 91
Stopper/Gate 92
Sensor Devices 93
Mistake Proof Your Mistake Proofing Device 93
Setup Reduction or Quick Changeover 93

7 Functional Applications of Lean 97
Sales and Marketing 98
Stability Issues with Sales and Marketing 99
Standardizing Sales and Marketing Processes 100
Making the Sales and Marketing Function Visible 102
Improving the Sales and Marketing Function 103
Purchasing 105
Stability Issues Relating to the Purchasing Function 106
Standardizing Purchasing Processes 107
Making the Purchasing Function Visual 108
Improving the Purchasing Function 109
Accounting 109
Stability Issues in the Accounting Function 110
Standardizing Accounting Processes 110
Making the Accounting Function Visual 114
Improving the Accounting Function 115
Customer Service 116
Stability Issues with Customer Service 116
Standardizing Customer Service Processes 118
Making Customer Service Visual 119
Improving the Customer Service Function 120

Human Resources 122
Stability Issues Relating to Human Resources 123
Standardizing HR Processes 124
Making the HR Function Visible 125
Improving the HR Function 126
Summary 128

8 Leading the Lean Organization 129
Driving Continuous Improvement (PDCA) 131
Mentoring 134
Going to the Gemba 138
Performance Measurement 139
Recognition 141
Summary 141

The Quality Toolbox 143

Forms 149

Index 163

Introduction

Lean thinking is all about common sense: creating a culture of continuous improvement in which all members of the organization actively work to improve business performance over time. Every Lean enterprise seeks to maximize the value delivered to its customers while minimizing waste. Lean organizations do this by optimizing the flow of products and services, at the pull of the customer demand. It is hard to imagine a business, regardless of industry, that *would not* want to maximize value while minimizing waste. Can you think of any organizations that would not want such a business model?

And therein lies the challenge. Lean cannot be a stand-alone initiative. Your goal is to make it a true business model—to make it *the* way of conducting business. Ultimately, you need to apply Lean to all aspects of your business, from the development and deployment of your business strategy, to the development of your organization's human assets. You need to apply it throughout all business processes, from obtaining the "voice of the customer" to understanding customer needs, to the delivery of products and services. Only then can you sustain Lean practices over the years and throughout changes in management.

It is a lofty goal, but that should not discourage you from embarking or continuing on the Lean path. Unfortunately, many organizations look at the tremendous success of companies who have gone before and fail to recognize the road that they have traveled. Remember that the goal is really incremental continuous improvement, and do not be discouraged as long as you are making progress. The good news is that you can learn from others and significantly reduce your learning curve—though not overnight, and not in a few years. You need five to ten years for Lean to sufficiently take root and become sustainable. There are no guarantees, but your chances of success are much greater if you realize that you are in it for the long haul. Lean is not a program, because a program, by definition, has a beginning and an end.

There are many well-documented Lean successes in manufacturing applications, and countless Lean manufacturing books. However, service organizations and administrative processes within manufacturing companies often struggle to apply these concepts. While there is an ever-growing number of books on "Lean office" and "Lean service," most do not treat the subject in sufficient depth, focusing strictly on the "tools" but not on their application. In turn, what is put into practice at most companies falls far short of Lean's lofty goals. Organizations fail to fundamentally change how work is performed and flows, and therefore do not realize the significant results that can be achieved. It is discouraging, and can lead to the abandonment of Lean altogether. Staff members then view the effort as "just another program"—and a failed one, at that.

Another pitfall of many Lean office and service efforts is a lack of alignment to organizational strategy and business objectives. When you learn a new tool, it is natural to want to immediately apply it. But these well-intentioned efforts do not always provide the expected results, because they are not aligned with the key objectives of the business or do not address a key business need. Management then becomes discouraged, and once again the effort is abandoned. This is not because Lean concepts are ineffective; rather, management did not thoughtfully consider the particular business processes that needed to be redesigned for the organization to realize its Lean objectives. Alignment is critical.

Your Lean office and service effort should focus on the key business processes that directly affect your ability to deliver value to your customers, such as order processing. Your business can realize near-immediate benefits, and your customers will quickly see the results. But there are other processes that you can address, for other business reasons. At one company, the new-hire process required immediate attention, as there was a pressing need to effectively and efficiently employ new associates to meet the increasing demand. At another company, an urgent need for cash required a focus on processes that would help generate working capital, including accounting, such as supplier returns and rebates, customer invoice disputes and resolution. Each organization needs to develop a set of clear objectives, identify the related business processes, and determine which to address.

When you focus *only* on tools, you are more likely to make isolated changes within departments and functions without changing the fundamental way that work is performed and flows. A tool is only as good as its application, and you need to apply Lean tools within the context of the overall business process redesign to realize the full benefits. Fortunately, there is a tool, Value Stream Mapping, to help you avoid this problem. Value Stream Mapping can help you see the overall process and the big picture. Although I will reference Value Stream Mapping, the assessment and planning tool of Lean practitioners, in this book, I will not cover it in great detail. For the definitive resource on Value Stream Mapping, pick up a copy of *The Complete Lean Enterprise: Value Stream Mapping for Office and Administrative Processes* (Productivity Press, 2004). For now, it is enough to know that using it properly will ensure that a Lean office and service effort will be more effective.

In other companies, the "Lean office" or "Lean service" effort is confined to trying to better organize the workplace through "5S" techniques (we will talk about these more in Chapter 6). These organizations fail to implement the key Lean concepts of standard work, flow, and level pull. Why do so many companies struggle with the application of commonsense Lean concepts to office and service processes? The stock reason is that the nature of work performed in the office or service is "different." How is it different? Most people point to the variability of the work, the multitasking, the unpredictability of demand, and the "creative" nature of the work. And it is true: work performed in the office and in services does tend to be highly variable. However, this occurs for several reasons, most of which are created by the companies themselves. For example, the lack of standard work, and its continued acceptance by management, is a primary cause for variability in office and services. The lack of standard work often results in inconsistent information quality, which requires extended time to address and correct. Another source of demand variability is batch processing, or performing an activity periodically, often for the convenience of the staff member. Individuals do not realize that they are batching and the impact it has on downstream processes. The good news is that you can reduce variability through the application of Lean concepts.

There are four basic steps to the application of Lean:

- Stabilize
- Standardize
- Visualize
- Continually improve

The approach you will need to take depends on your starting point. If your process is highly unstable, in that it has an inconsistent and often unacceptable output, you will first want to *stabilize* it. If your process is already stable, though, you can start to *standardize* the way in which your processes are performed—in other words, identify and agree upon the best way to perform each activity or process and ensure that different people perform the same process consistently. Once your process has been standardized, you need to provide *visibility* to the process within the organization. And of course, your ultimate goal is to *improve* all processes on a continuous basis.

Next, we will explore each step in more detail. But first, a caution: the importance of accurately identifying your starting point cannot be overemphasized. For instance, if you attempt to standardize an unstable process, you are not likely to see the results you are hoping for.

Let us take a look at "Stabilize" first.

Stabilize

The objective of this step is to create predictable and repeatable outputs. In office and service environments, "we never know what we're going to get" is a common mantra. But this is not the case with manufacturing processes; even when they are not as efficient as they could be, in the end there is an assurance that the product will perform as the customer expects. So what's the difference? In manufacturing environments, inspection and test operations provide this assurance. However, in office and service environments, the "product" is not as tangible, so it can be more difficult to ensure the quality of the output. In service environments, in particular, the near-real-time creation and consumption of a service can make quality assurance problematic.

If your office or service process is incapable of delivering a consistent output, then your Lean effort must begin here. You need to identify the source of the instability, which most often is an inadequate understanding of customer needs. In rare cases, you may even find a complete disregard for meeting customer needs. "The customer will get it when we say so." Although you may not hear someone say it outright, the same meaning can be communicated by the continual poor performance of a given organization, department, or individual. In other words, service providers no longer hear the "voice of the customer."

In these situations, you will start fairly small, by clearly defining the needs of the customer, documenting them in simple ways (think checklists), and providing training to office and service personnel. Interestingly, implementing pull systems can provide much-needed stability. One of the objectives of pull systems is to give customers what they want, when they want it. Clearly, just achieving this objective on some consistent basis will bring stability to your process or system. That is one reason that implementing pull systems can be an early focus of a Lean office and service implementation. We will talk about this in more detail in Chapter 4.

What are some other causes of instability? Let us examine what happens when you have a poorly defined existing process. In this instance, your staff members are left on their own to figure out how best to perform a particular process. Let us further assume that the details of how this process is performed are restricted to specific persons (an arrangement often referred to as *tribal knowledge*) and are not adequately shared with others in the organization. Can you see how this could cause instability, since the outputs can greatly vary based on who performs the process? The fix, again, is fairly simple: you need to define a specific process that all employees can understand and follow. Mapping the entire process can provide much-needed definition. This leads us to our next Lean step, "Standardize."

Standardize

When implementing Lean in office and services, you will often find that you already have some stability within your processes, and standardizing will then become your starting point. When we standardize, we develop practices consistently followed by all people who perform the process and/or the activities linked to the process. Why do you care *how* I do it, as long as I get it done? It matters, and we will go into more detail about *why* it matters in Chapter 2. For now, know that in most cases there is some semblance of stability that can be further improved by standardizing. One primary focus of standardizing is to streamline or simplify work. Let us say that you have a process that takes 15 minutes, and by standardizing it, you are able to lessen its duration to 10, or even 5 minutes. Less variability will creep into the process strictly due to the shorter duration, and your employees are more likely to adhere to the process if they are known that it represents the best-known simplest way possible at the time it was developed.

As we will discuss later in the book, another purpose of standardizing is to make it easier to identify nonstandard conditions. These are conditions that must be addressed to return the process to acceptable levels of performance. Nonstandard conditions will not be recognizable if there are no standards to compare them with. For example, if everyone performs a particular activity in widely varying ways, what kinds of conclusions can you draw about efficiency and the impact on the customer process? By standardizing, you can make the answers to these basic questions readily apparent, and that leads us to our next step.

Make Visible

Our key objective with this step is to have the workplace "speak" to us. Visual communication is the most effective and efficient method of communication. Lean enterprises always look to improve visibility throughout their operations. Some people refer to this as "transparency." In the beginning, organizations will make performance more visible. This is a good first step. However, much more can be accomplished, but only once your processes and activities have been standardized.

A visual workplace is one that is easier to manage over time. With work instructions and prioritization rules posted visibly, less time will be needed to direct the most basic activities. With techniques that make performance more visible, less time will be needed to identify problems and issues with performance. The subject of visual management will be covered in detail in Chapter 5. For now, it is sufficient to say that a visual workplace makes it easier to drive continuous improvement, the real objective of Lean, and the next step.

Continually Improve

When you begin your Lean implementation, you will find no shortage of opportunities to improve. You will engage your people to improve performance. You may choose to begin with changes on a smaller scale, within existing departments or functions. You will achieve local improvements first, which will lay a foundation for broader changes to be made in the future. Alternately, you might choose to start off with the redesign of entire value streams. In other words, you will choose an approach that works for your situation.

Regardless of the approach, there will be a high level of improvement activity for some time, perhaps two to three years. But how will you sustain continuous improvement? How will continuous improvement become part of your organization's culture? Surely the visual management techniques to be implemented will greatly help in this regard. But you need more: you need effective leaders who provide a learning environment that is safe for experimentation. *And* you need personnel development practices to sustain the system even in the event of a change in leadership.

Few organizations sufficiently invest in the development of their people. There are companies that have successfully applied Lean for many years, 5 or even 10 years, but still "lost their way." Often, this is attributed to the failure to continually develop leaders who deeply understand Lean, who can sustain and even improve the system, and who can teach it to others. Only in this way can the culture of continuous improvement be sustained. As commonsensical as Lean is, it is still not common practice.

Getting Started

As you read the following chapters, you'll need to consider the concepts covered in this book in the context of these four steps: stabilize, standardize, visualize, and improve. Again, the approach you take will depend on your starting point.

In this book, we will go beyond the basic tools and delve into the key concepts of Lean as they apply to office and services. First, we will discuss Value Stream Management, followed by chapters on Standard Work, Flow, Level Pull, and Visual Management. Next, we will cover several of Lean's more important tools, such as 5S and Mistake Proofing. You will apply these while implementing the key concepts covered in Chapters 1 through 5 in order to maximize the benefits.

Chapter 7 provides a function-by-function review of the application of Lean. Does Lean apply to Sales and Marketing? How about Accounting? We will cover the functions commonly found in most service organizations and the administrative areas of manufacturing companies. The examples provided in this chapter must be implemented in the context of Value Stream Management in order to maximize the benefits.

While the commonsensical nature of Lean concepts will resonate with most people, the successful application of Lean requires fundamental behavioral change in many people. People are creatures of habit, and have difficulty changing. But you can create *new* habits, given sufficient time and support. The most common obstacles encountered will be addressed throughout the book, along with strategies to overcome them, drawing on over 20 years of practical experience. This discussion will culminate in Chapter 8. By then, you will have confidence to put into practice these well-tested and proved methodologies.

Let's get started.

Chapter 1

Organizing by Value Stream

In this chapter, we will look at three alternatives for organizing by value stream:

- Creating cross-functional or intradepartmental, colocated teams based on value stream
- Defining roles based on value stream within a department
- Organizing activities for individuals who are supporting multiple value streams

The approach you will need to use will depend on the complexity of the overall process, the number of people involved, and the level of knowledge and skill required to perform these necessary activities, among other considerations. But first, some background.

The *Lean Lexicon* (Lean Enterprise Institute, Fourth Edition, March 2008) defines a value stream as "the set of all actions, both value-creating and nonvalue creating, required to bring a product (or service) from concept to launch (also known as the development value stream), and from order to delivery (also know as the operational value stream). These include actions to process information from the customer and actions to transform the product (or service) on its way to the customer."

There are also a number of information processes that businesses perform to support these two primary value streams. These "secondary" information processes often do not create value for the external customer, but are necessary to run the business. For example, accounting activities in a manufacturing company are performed in support of the primary operational value stream (and purpose) of the business: to make and deliver product.

However, for an accounting firm that is providing accounting services to another company, it is its primary operational value stream.

One of the most formidable obstacles to applying Lean Thinking to office and service processes is the existing organizational structure. Most companies are organized by functions or departments that have little relationship to the information actually being processed or the service being delivered. In most cases, these functional structures impede the flow of information. They can result in parochialism, where managers may be responsible for portions of the information or service flow, but where nobody is responsible for the management and continuous improvement of the overall process. More specifically, they tend to add to the number of hand-offs and often give rise to conflicts of priorities between departments. Budgetary practices can further increase the likelihood and effects of these conflicts, as "turf battles" occur between departments as they compete for limited resources and funds.

We have seen this in manufacturing for years, where functional organizations existed with separate machining, assembly, quality assurance, packaging, and shipping departments. This approach was thought to optimize individual functions' performance, though this often came at the expense of overall process or system performance. Although many manufacturing organizations have realized that they must change the fundamental ways in which they are organized to realize breakthrough results in improved material and product flow, office and service environments are just starting to come to the same conclusion.

In either environment, your objective is the same: to optimize the performance of the overall value stream. In an office or service environment, this means optimizing the performance of each information or service process. You will also need to ensure the ongoing continuous improvement of your business's primary value streams and secondary processes. You will do this by organizing and managing by value streams, or what some have called "end-to-end" process management. Ideally, instead of having different managers responsible for particular processes within the same value stream, a single person—a Value Stream Manager will have the overall responsibility for a specific value stream's performance. The point is not to have a Customer Service manager focused on the Order Entry activity and an Accounting manager focusing on the Invoicing activity, and so on. The real objective is to ensure that all the activities in the value stream are being performed in ways that optimize the performance of the overall "end-to-end" process. Consider an organizational structure based on actual information flows: "order-to-cash," "requisition-to-pay," "product and process development," etc. What advantages will your organization realize by having order processing, planning, invoicing, and collection personnel—an "order-to-cash" team—working *together* as a team rather than as separate departments? What if this team reports to a single value stream manager, with performance metrics based on the overall process? What are the disadvantages of this organizational structure?

What I have just described can be considered an ideal. But organizations do not really need to change their structure and reporting relationships in order to improve the performance of processes or value streams that involve multiple departments. There are other alternatives that can provide meaningful results. In the following sections, we will review the three ways to organize by value stream and the advantages and disadvantages of each. For each method, existing reporting relationships can remain the same and can remain effective. First, let us look at cross-functional teams based on value stream.

Cross-Functional Teams Based on Value Stream

More and more organizations have set up office "cells" or "pods"—cross-functional teams of people, colocated to process information in fractions of time. Office cells can reduce lead time, including queue time, by as much as 90%, and process time, or "touch time," by as much as 40%. Office "cells" can improve the effectiveness and efficiency of communication and decision making, while reducing quality "defect" or "correction" waste. The number of hand-offs can be reduced, and a greater awareness of the needs of "internal" customers can be achieved. The result is the reduction of non-value-added waste throughout the value stream. In Chapter 3, I will show you how to implement cross-functional teams. For now, a general discussion of the nature of the team structure (which will depend on the information being processed or the services being delivered) and the obstacles to establishing it will suffice.

How about secondary processes that are only performed periodically, such as payroll, month-end closing (closing the financial "books" at the end of each month), generating proposals, and similar processes? Can cross-functional, colocated teams be set up to perform these processes and achieve similar results? Definitely. Consider what most companies do when they are working in a compressed timeframe: they typically identify a cross-functional team to meet the challenge. And often, the team will take on the

Figure 1.1 Functional versus cell approaches.

challenge while physically occupying the same space, a conference or meeting room.

If it makes sense to organize ourselves in this way when we are in a hurry, why not organize ourselves this way whenever the opportunity arises? A typical objection is that it is disruptive. But it is only disruptive because we do not typically plan for it. If you organize your work structures to support this approach from the outset, it need not be as disruptive as you might expect. Further, if the task at hand is completed in significantly shorter timeframes and with less, not more effort (remember the typical results in shorter lead and process times), then people can return to their "normal" duties sooner and have *more* time for them as a result. Sound like a win-win? Take a look at Figure 1.1 to see how you might organize your cell approach. The figure on the left side depicts a traditional approach with separate departments located in different areas of an office or building. The figure on the right side shows a cross-functional team physically located in the same area.

Defining Roles by Value Stream within a Department

We can apply the same concept of organizing by information process at the department level. One breakthrough exercise that can help people begin to see how to apply Lean is to identify "service families." When you identify service families, you begin to identify the key services delivered by a department or function. You will begin to see the "processes" that are being regularly performed, and realize that they are *not* adequately organized by service or process.

Let us take a look at Customer Service. Customer service personnel are often involved in multiple activities that fall under the umbrella function of

HOURS OF OPERATION / PHONE COVERAGE 7:30 - 5:00

PHONES / TASK | ORDER ENTRY

SCHEDULE	WHO	HOURS	7:30	8:00	8:30	9:00	9:30	10:00	10:30	11:00	11:30	12:00	12:30	1:00	1:30	2:00	2:30	3:00	3:30	4:00	4:30
A	Keith Jesse Dan	7:30 to 4:30								LUNCH	LUNCH										
B	Armand Steven Leslie	8:00 to 5:00									LUNCH	LUNCH									
C	Nicole Mark Allson	8:00 to 5:00											LUNCH	LUNCH							

Figure 1.2 Organizing activities for a department by value stream.

"customer service." Personnel are expected to determine the best way to organize themselves; in other words, to work in a way that works best for the individual. The result is a very unpredictable, highly variable work environment with little or no standard work throughout.

An alternative is to organize by service family. There are typically three key services delivered by customer service personnel: order processing, problem solving (e.g., providing technical support, order status, or order maintenance), and value creation (e.g., proactively generating sales). Unfortunately, so much time is spent on the first two that little time is left to create more value for the business. But what if the personnel within the Customer Service department were organized by these three service families? How would this lend itself to greater organization of the activities? How would that impact the effectiveness and efficiency of these activities? Would it reduce variability and help achieve standard work?

When you require that particular people within the customer service department perform particular activities and have separate roles and responsibilities that *does not* mean that it is all that these people will do forever. Maintaining an enriching work environment is important in a Lean enterprise, and rotation will prevent people from falling into the trap of performing mind-numbing repetitive work over time. I encourage you to rotate people between roles in a way that will minimize disruption in the operation. Let us take a look at how this might work in a Customer Service department (see Figure 1.2).

In this case, two different categories of activities were identified: "Order Management" and "Task Management." Task Management involved highly variable and unpredictable activities, such as answering inbound phone calls, logging interactions, responding to e-mail, and similar tasks. Order Management activities included entering orders, processing return material authorizations, and the like. These two categories were identified to separate the highly variable activities from the more standard activities. There is *no* reason to allow standard activities to be negatively impacted by the highly variable ones. Of course, maintaining telephone coverage was vital, so management performed a study of inbound calls and established a schedule for the department accordingly. Customer Service associates were assigned to one of three schedules, each with designated times for Order Management activities and separate times for Task Management activities. This schedule ensured that sufficient capacity existed to answer inbound calls in a timely manner throughout the day. As a result, the department was able to reduce order entry errors and increase throughput. Equally important, the associates felt that the work environment was much more productive and satisfying.

Organizing Activities for Individuals by Value Stream

Even if the same resources are expected to perform multiple tasks, how can we better organize the completion of these tasks? Too often, office personnel are left on their own to decide how and when to perform the multiple tasks expected of them. Unfortunately, what that means is that each person in the office usually organizes him- or herself in different ways. Why is that a problem? First, the timing of the completion of work may not be appropriate. Two, the prioritization of work is likely inconsistent through the value stream. Once again, this contributes to a standard office problem: the unpredictability of the movement of work from one step to the next.

What if a "plan for every process" was developed for all of the key information processes in the office? With a plan for every process, each task will be assigned to specific days of the week and times of the day. Further, the plans for different people (i.e., roles) in the office will be "synchronized" to each other to maximize the flow of information and service. This way, we can guarantee tremendous predictability in the multitasking work environment. Each person will know what to do and when. We can avoid simultaneous processing of different information by the same person, another way to provide significant productivity improvements. It is typically easier for

Time	Monday	Tuesday
9:00–9:30 a.m.	Check e-mail	Check e-mail
9:30–10:00 a.m.	Enter orders	Enter orders
10:00–10:30 a.m.		
10:30–11:00 a.m.	Unscheduled work	Unscheduled work
11:00–11:30 a.m.	Process RMAs	Work on order holds
Etc.	Etc.	Etc.

Figure 1.3 An example of a "plan for every process."

someone to perform one task at a time than attempting to do three things at once. In our Customer Service example, trying to answer the phone while entering an order can give rise to possible errors, as well as increase process time. Can the number of disruptions and interruptions be minimized by developing a "plan for every process"? This is analogous to the "plan for every part" concept applied in manufacturing, in which delivery times and quantities are set for every part used in the manufacturing process. Let us take a look at what a plan for every process might look like (see Figure 1.3).

Even time for "unscheduled work" has been put aside. This time can be used for unplanned or "drop-in" work. The details of how this might work will depend on the nature of the individual's position, among other factors such as the urgency of the unplanned work, and the true response time required. Of course, you will always need to maintain satisfactory service to internal and external customers. Nevertheless, such approaches have been proved effective in Human Resources, Information Systems, Customer Service, and other functions. For example, a human resource professional may want to establish designated hours for employees to visit to discuss personnel issues rather than allowing employees to drop by randomly.

Summary

The way in which you organize (or do not organize) your activities can contribute to the very problems that might make you believe that Lean is *not* applicable to the office or service environment. Remember, the office is not different, and organizing by value stream is not just possible—it is *critical.* In the following chapters, I will show you how it is done.

Three alternatives for organizing by value stream:

- Creating cross-functional or intradepartmental, colocated teams based on value stream
- Defining roles based on value stream within a department
- Organizing activities for individuals who are supporting multiple value streams

Chapter 2

Creating Standard Work for Office and Service

You may have heard that standard work is one of the foundations of Lean Thinking. However, developing and practicing standard work can be elusive, particularly in an office and service setting. What are the difficulties? Why does this very basic concept so often generate strong resistance? Most often, the answers to these questions lie in a lack of understanding of standard work and its benefits. In general, we resist what we do not understand, and we are unlikely to change our ways if we do not see any benefit to doing so.

Standard work is a critical concept, and in this chapter, we will explore it in depth, including why you should implement it, its benefits, and its necessary elements. In the process, I will provide you with possible responses to any naysayers. Typical arguments against standard work include

- "Why do you care *how* I perform my duties, as long as I get the work done?"
- "Standard work *does not* work for the 'creative' activities we perform in office and services."
- "The office (or service) environment is too variable. It does not lend itself to standard work."

I will debunk each of these myths and, in the process, show you how to effectively implement standard work for most all activities performed in *any* office or service environment.

Standard Work—What It Is and Its Purpose

Simply put, standard work is the best-known way to effectively and efficiently perform an activity. Standard work defines the desired sequence of steps, the time required to perform the steps, and other elements that ensure that an activity is performed in a consistent way over time. Not only does this ensure that the process or activity itself is performed consistently, it ensures consistent quality of the "output" of the process. Standard work is intended to be used in conjunction with, but *not* in place of, training.

Documentation of standard work must be simple and visual; in other words, it needs to actually be posted in the area where the work is performed. Do not misunderstand me: this does not mean that anyone should be able to pick up standard work documentation and be able to perform the activity that it covers. There is a distinction between standard work and "standard operating procedures," or SOPs. SOPs are detailed work procedures that can be helpful for new employees or employees who are new to a particular activity or process. But SOPs, though good reference tools during a new employee's learning curve, are not replacements for standard work. So, what is the difference between a SOP and standard work? Well, standard work does display the "what to do" and, to some degree, the "how to do it," but typically *not* at the level of detail that a SOP would. Further, standard work instructions are to be used as a reference for someone who has *already* been trained so that consistent practices can be maintained.

A typical challenge when implementing standard work is striking the balance between too much detail and not enough. You need to group specific steps together and describe them in ways that *remind* people what to do without necessarily *spelling out* what to do. For example, if you are implementing standard work for an Order Entry process, you do not want to include details such as "log onto computer, go to screen ABC, enter field #1, go to field #2," etc. Sure, those details are helpful for new employees or employees switching work responsibilities, and you can supplement your standard work with a detailed procedure or SOP to provide those specifics to be referred to as needed. You need to determine the appropriate level of detail for standard work, and that comes with practice—and with a better understanding of the purpose of standard work.

One purpose of standard work is to identify when nonstandard conditions arise and, in turn, trigger an action to correct or improve on those conditions. However, we cannot identify nonstandard conditions if we do not have standards to begin with. Nonstandard conditions can include

- Failure to perform an activity
- Failure to perform an activity at a required point in time
- Taking longer to perform an activity than it should
- Performing an activity in a way that will have a negative impact on some "downstream" process

We need to quickly identify these situations, since they represent opportunities to return the process back to its desired standard condition. Ideally, the people performing the activity can identify these conditions, self-correct the process, and maintain standard work over time. Short of this ideal, others can observe whether or not the members of an organization are adhering to established standard work practices. Lean enterprises encourage managers to conduct periodic process observations—not in any kind of punitive sense, but because opportunities for improvement can be identified in this way. We will talk more about process observation activities in Chapter 8.

In a Lean enterprise, it is the responsibility of leaders (e.g., supervisors, managers) to ensure that standard work is practiced throughout their areas of responsibility. When employees are not following established standard work practices, the leader has an opportunity to affect process improvement. For example, the leader may train an employee to improve his or her skills so that an activity can be performed within the expected timeframe. However, if each employee is allowed to perform activities in substantially different ways, variability runs rampant. A lot of the variability within an office or service environment comes from a lack of standard work and, with that lack of standard work, it becomes difficult for leaders to ascertain how both to correct and how to reward employees. This variability is pointed to as the reason that "Lean does not work in the office or service environment." But this is really a chicken-and-egg argument, since standard work and the manner in which employees perform their tasks can actually *reduce* variability within office and service environments. Before we get further into the benefits of standard work, though let us take a look at the elements that make up standard work. These include the specific tasks that are to be performed, the "key points," and the time and timing of the tasks.

Elements of Standard Work

The "What"

The first element of standard work is the "what"—in other words, defining the tasks to be performed. As we touched on briefly earlier, specific steps need to be grouped and described, and those groupings listed, in an effective and efficient sequence. As we begin to get into the "how," we cross over into the "key points"—the information employees need to adequately perform steps within a process.

Key Points—The "How" and "Why"

You can think of key points as the "how" to perform a step within a process. Often, key points comprise "tribal knowledge," or knowledge that people possess but do not adequately share with others and that is required to correctly perform a task. Often, such important details are never adequately documented and are lost over time, as people change responsibilities. Again, when we are describing *how* to perform a process, we need to find an appropriate level of detail—not too much, not too little. Remember that standard work is not a training tool, but a reference for somebody who has *already* been trained, even if that person has not performed the task in question within a reasonable period of time (e.g., a day, a week, or a month). In this instance, standard work should enable that employee to pick up the standard work documentation and quickly refamiliarize him or herself and perform the task in an effective way, by providing the correct output. It is only natural that this employee may need a few iterations to perform it efficiently, but we should *immediately* expect the correct output.

Key points tend to relate to quality, efficiency, and safety. Your key points will include the details needed to guarantee an acceptable quality result. Details regarding the fastest way to perform a task (while assuring quality) are provided. When appropriate, key points should also include details that ensure that an activity is performed safely. For an Order Entry activity, a key point for a particular step may be to "correctly input all fields identified as required." Otherwise, the order will not be accurately processed. This begins to provide a "why," an important element of standard work.

Why do people drift away from standard work procedures over time? Let us say that an employee performing a task discovers a change that seems to help himself—a shortcut, maybe, that enables him to complete his work more quickly. Will he be likely to play the scenario out and consider whether his change creates an unintended consequence further down the line? What if this change causes quality problems for a downstream process? The bottom line is that well-intentioned employees taking shortcuts can create new problems. However, the "why" can help us avoid that explaining the logic behind the defined tasks, how they are sequenced, and the manner in which they should be performed. People are more likely to maintain standard work procedures if they understand the reason or reasons behind it. The "why" is critical. Include it in standard work.

Time and Timing

Standard work also includes the expected time to complete a task or group of tasks. Let us say that the expected time for entering an order is 5 minutes. If an employee is regularly taking 15 minutes to enter orders, that is a nonstandard condition. Ideally, the employee should point this out to his or her manager. Potential causes include a lack of training or unforeseen circumstances that may have crept into the process.

Often, organizations are reluctant to include process time in standard work. It is an understandable concern, particularly given that failure to meet expected times can result in punitive responses in traditional work environments. But rest easy: in a Lean enterprise, identifying the failure to meet expected times represents an opportunity for improvement. Further, expected process times can be expressed as a range to accommodate the inevitable variability in some office and service processes. You will need to agree upon an acceptable range within your organization. Process times beyond that range will likely identify a nonstandard condition that might need to be acted upon.

Timing of tasks can be important in an office and service process; in other words, a task may need to be completed at a specific time of the day, or day of the week to ensure that other tasks can be performed by other departments or functions in a timely and/or accurate manner. This kind of detail can be lost over time as people change roles and responsibilities, so you need to include it in standard work when appropriate.

Let us look at how we can document standard work simply and effectively, while ensuring that all of the elements discussed are present.

Sidebar: Including Demand Rate in Standard Work

Readers who have previous experience in standard work may have noticed that demand rate is not included here as an element. That is because it is not always possible to determine a meaningful rate of demand for an office or service process. For example, a hiring process may be performed periodically, on an as-needed basis. You can still develop standard work for this process but without the demand rate element. Nevertheless, an example of how demand rate is incorporated in processes with more predictable demand appears in Chapter 3, "Creating Flow in Office and Services."

Visually Displaying Standard Work

Out of sight really can be out of mind—so you need to visually display your standard work for a process or a step within a process, ideally on one page. For certain processes, this can turn into two pages, but your goal has to be to avoid "books" of procedures. Figure 2.1 illustrates a successful format.

As you can see in Figure 2.1, all of the key elements of standard work are represented: identification of the task, the key points that relate to a task,

Standard Work Instruction			
Process: __Order Processing________________ Operation: _Order Entry___________________			
Task	**Key Points**	**Time / Timing**	**Visual References**
1. Enter Order	- Enter header first, then each line item for efficiency - All required fields must be input to insure accuracy	- 5-10 mins per order - Enter within day of receipt	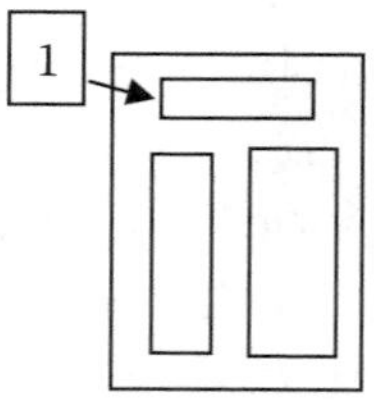 1
Etc.	Etc.	Etc.	

Figure 2.1 Example of standard work instruction.

and the expected time or timing of the task. In addition, we have included a space for visual references that can be used to clarify tasks and/or key points. A picture can be worth a thousand words when it adds important detail without adding complexity to the documentation. Using our Order Entry process, for example, "screen saves" can be captured and included in the document. Perhaps, a process flowchart will help to add clarity regarding the general flow of work.

Let us take a look at another format that is effective for a multitasking environment (see Figure 2.2).

This format illustrates the list of tasks that a Customer Service employee might perform, along with the time required to perform the task and the desired schedule. Key points clarify the purpose of the task. This format can be easily expanded to provide clarification on *how* to perform the task, including visual references. If done right, any person expected to fulfill the role of Customer Service would know which tasks are to be performed and when, how long it should take to perform each, and have some guidance as to how and why to perform each task. The employee will still need training to perform each task, but there is an inherent benefit in having a document to support the training at the ready.

Next, let us take a look at the benefits of standard work.

Standard Work Daily Management				
Role: Customer Service				
		Frequency		
Task (with key points)	**Time**	**Daily**	**Weekly**	**Monthly**
1. Enter Orders within day of receipt to ensure that published lead times can be met	5-10 mins per order	Throughout day		
2. Generate weekly order input reports to monitor current demand	5 mins		Fridays by 3:00 PM	
3. Generate monthly reports for management to monitor sales performance	10 mins			Last Friday of month

Figure 2.2 Example of standard work in a multitasking environment.

Benefits

From personal experience, applying standard work to office and service processes can result in a number of benefits:

- Learning curves are reduced by up to 75%, which also makes it easier to cross-train employees
- Productivity or efficiency improvements of 10% to 25%
- Greater flexibility through cross-training to respond to changes in demand and to better accommodate staffing changes (e.g., absenteeism, turnover)—even for "creative work" (marketing, design).
- Maintaining and even improving customer service and satisfaction, as the outputs of a process (e.g., order delivery time) become more consistent.

When coupled with the time-tested teaching technique of Job Instruction (JI), the reduction in learning curve attributed to the application of standard work is up to 75%, and has been well documented for over 60 years (*Training Within Industry: The Foundation of Lean* by Donald Dinero, Productivity Press, 2005). The productivity or efficiency improvement that results from the development of standard work stems from the collaborative effort to identify "best practices." The streamlining of current activities is a part of the process of developing standard work. It is not simply a documentation effort.

SIDEBAR Job Instruction (JI)

Job Instruction (JI) is a technique to teach standard work to others. It was fully developed during WWII as part of an effort called "Training Within Industries (TWI)," though its true origins date back to WWI. During WWII, realizing that there was a surfeit of skilled resources in key industries to support the war effort, the U.S. War Department assembled a team of individuals (who originally pioneered these training techniques during WWI) to teach companies an effective way to train people in shorter timeframes. The time to train was reduced by as much as 75% by application of the simple methods defined in JI (we will go into these in more detail in Chapter 8). Unfortunately, the technique was all but forgotten in the United States after WWII, but in Japan, Toyota did learn the concepts and has been consistently applying them ever since the 1950s. It is a foundation concept of Lean.

As learning curves are reduced and work is simplified, you will notice a substantial increase in flexibility. This is particularly true of what has been called "creative work," such as design, engineering, and marketing activities. A frequent argument is that standard work *cannot* be developed for "creative work." However, such processes involve mostly information flow—identifying what information is needed, where to obtain it, and what to do with it. That is a process, and *all* processes lend themselves to standard work. The truly creative portion typically represents a small percentage (< 15%) of the work content, making this argument largely unfounded.

Steps to Creating Standard Work

Now that you have learned about the purpose of standard work, its elements, and the formats that it can take (as shown in Figures 2.1 and 2.2), we will review the general approach to creating standard work. But first, a caution: one common pitfall of the application of standard work is to attempt to create it for every activity that people perform. This just is not practical, and it can send an unintended message to the organization that Lean is about controlling everything that people do. Instead, standard work must be applied to the key activities performed that ultimately affect the customer and the overall performance of an important business process. Always ask this question: "How will the individual and/or the organization benefit by the creation of standard work for this activity?" In short, use common sense when applying any Lean concepts.

The steps to creating standard work in an office and service environment are

1. Identify the key activities that are performed in an area
2. Prioritize these by importance (optional)
3. For each key activity, identify a team of individuals who will develop the standard work
4. Observe the current process, identify differences between associates and opportunities to streamline
5. Obtain consensus on "best practices"
6. Document it in a simple and visual way
7. Train associates in the new standard work
8. Monitor for effectiveness, issues, and compliance

During Step 1, you need to identify *all* activities performed in a particular area or department. However, standard work should focus only on key activities. Therefore, you will want to prioritize your identified activities, in order of importance, during Step 2. For example, a single role, such as Inside Sales, may involve ten or more activities in the course of a day or week. It may not be practical or necessary to develop standard work for each of these, but common sense dictates that you should focus on the activities that constitute the largest percentages of people's time. To determine which activities to prioritize for standard work, you may need to conduct some data collection with the people performing the activities. Do this before the standard work development effort begins.

In Step 3, you will assemble the team that will develop the standard work. Obviously, the team must include associates who perform the activity on a regular basis. It may not be possible to include everyone from an area on the team; in this instance, the associates selected for the team must understand that they represent everyone's interests. Nonteam members will be brought into the process periodically at appropriate times, in particular, during Steps 4 and 5. A leader, supervisor, or manager can be part of the team, but he or she must remember not to usurp the process. The leader, supervisor, or manager can step in the situation that consensus is not possible in Step 5. An "outside set of eyes" is always welcomed, since people from outside the area tend to ask excellent questions about the how, what, and why. A resource from the information technology (IT) group is always welcome, since IT representatives can often identify opportunities to streamline activities. Altogether, your team will typically consist of five to six individuals.

Step 4, observation of the current process, is very important. In Step 4, team members and nonteam members are observed to identify differences in current approaches and develop a consensus on "best practices." Step 4 also provides an opportunity for the associates to step out of their process and challenge both what they currently do, and how they do it. Almost invariably, Step 4 will help to identify opportunities to reduce the time that people currently need to perform an activity, and will surface non-value-adding (NVA) extra processing waste (one of the eight categories of waste; see sidebar), On some occasions, the opportunity to eliminate entire steps will be identified. By seeking ways to streamline, it sends the correct message that the development of standard work will help to make people's jobs easier. This promotes buy-in of the associates affected in the effort.

SIDEBAR The Eight Wastes

The reduction or elimination of non-value-added activities or waste is better understood today by those that have some experience in the application of Lean to office and services. True, people continue to choose to "explain them away" as necessary or "business-required," but this mind-set means that a lot of waste goes unchallenged and therefore unchanged. The eight categories are

1. Overproduction: producing more information or service than is needed and/or sooner than is needed by the customer. Examples: overly detailed reports, highly detailed planning processes, financial planning too far into the future.
2. Inventory: anything that is in excess of one "piece" flow, or "batch" processing. Examples: processing invoices once per week, completing all performance evaluations at the same time of the year.
3. Correction: any activity that is performed to correct an error. Examples: correcting order entry errors, issuing credits due to invoice errors, hiring an unqualified person for a position.

4. Extra processing: steps that take longer than they should, or entire activities that do not add value for the customer. Examples: a company's budgeting process (adds no value to the external customer, and questionable value to internal customers), meetings that take longer to conduct than they should.
5. Motion: movement of office and service personnel. Examples: walking to and from centralized files, printers, the facsimile machine, etc.
6. Transportation: movement of information or materials. Examples: e-mail, handing off paperwork.
7. Waiting: customers or information waiting to be serviced or worked on. Examples: waiting for decisions to be made, problems to be resolved, etc.
8. Underutilized people: not using people's full skills and abilities. Examples: providing people with narrow responsibilities, insufficient cross-training, overly limited authority.

The discovery process that represents Step 4 should lead to agreement on "best practices." However, obtaining consensus (see Step 5) is not always easy. Collecting additional data (see Step 4), including process time or quality information, may help here. If an associate has a faster way of performing a task, and the data backs it up, most people will concede that it should be part of the standard work.

The process really should be a collaboration, which in turn will garner greater commitment. To get that commitment, the team should review its recommendations with nonteam members during Step 5. In the absence of a consensus, the leader, supervisor, or manager can still ask people to commit to the recommendations, with the possibility for changes that may be identified as part of Step 8. In Step 6, you will document these best practices, perhaps using a format similar to the ones provided earlier in this chapter. Next, during Step 7, all associates who perform the activity will be trained. Ideally, this training should use the Job Instruction (JI) methodology previously mentioned.

Step 8 is critically important. Standard Work cannot be simply created and imposed on people without leadership or management follow-up to verify its effectiveness. He or she should spend time with the associates, observe, and listen. He or she must ensure that new problems did not arise as a result of the Standard Work created. These interactions can provide an opportunity to see if the training was effective and to reinforce the importance of the effort.

Summary

The fundamental concept of standard work is indeed achievable in an office and service environment. What's more, the benefits can even exceed those experienced in manufacturing as there tends to be little or no existing standard work in office and service. All that is required is the willingness to apply this critical concept, and to not allow the traditional arguments to prevail.

Standard work can be applied to any repetitive activity that any person performs in any department or function in any company. The key is to focus on activities, not titles. And "repetitive" is a relative term. It can be an activity that is performed once a year, such as the annual budgetary process—this too can (and should) be standardized. Therefore, it can apply to Sales and Marketing, Customer Service, Purchasing, Scheduling, Accounting—all support functions in manufacturers and nonmanufacturers alike. It can apply to how guests are greeted and processed in the hospitality industry, and how patients are processed in health care. It can apply to tellers at commercial banks and call center attendants and technical support centers. There is really no place that it *does not* apply.

The result of the implementation of standard work is a more predictable and stable process. This provides an excellent foundation for the application of flow concepts, which we will cover in the next chapter.

SIDEBAR Standardizing "Creative Work"

A company that we worked with creates marketing campaigns for consumer products. This particular organization argued that standard work, and Lean in general, does not apply due to the creative nature of the process. However, the company realized during a value stream mapping exercise that up to 90% of what they do is really a process—and a process that they very consistently follow, but failed to recognize. Market research, which is very process oriented and lends itself well to standardization, identifies the potential buyers of the products (e.g., males ages 18–49), as well as what media they frequent (e.g., radio, magazines). The only truly creative portion was the development of the radio spot, which accounted for approximately 15% of the total process time and 10% of the total lead time for the overall process. Once the company recognized this fact, they went forward in earnest to apply standard work to all of the noncreative processes and realized an overall lead time reduction of 50% and a process time reduction of 25%. The freed-up capacity and improved customer service allowed them to be more responsive to the ever-changing needs of the customers with regard to marketing efforts.

Still another example is in the Sales function, where people use the same argument of the creative nature of the sales activity. "Everyone sells differently. Every customer is different. You cannot have standard work for selling." People often confuse standard work with "style." We can still have a standard selling *process*, while salespeople utilize different "styles" that they feel is appropriate. A step in the sales process may be identified as "Establishing Rapport." One salesperson may choose to do this by asking questions about the family, while another engages in friendly banter about local sports.

Chapter 3

Creating Flow in Office and Services

The four key concepts of Lean as defined by Womack and Jones in *Lean Thinking* (1996) are value, flow, pull, and perfection. In this chapter, we will discuss in detail the application of flow to office and service processes. Here is what we will cover:

- The three possible approaches to flow, and the advantages and disadvantages of each
- A process for designing flow systems in offices and services
- The typical results you can expect when you apply this powerful concept

And once again, I will debunk common arguments against considering and implementing flow in offices and services.

In Lean terms, the ideal for flow is "one-piece flow," in which information is processed or service is delivered uninterrupted with minimal or no queues. This results in a shortened or eliminated wait time for performing a process or delivering a service, which usually means more satisfied customers. However, what is a "piece" in office and services? We need to define an appropriate "unit," understanding that this definition will vary depending on the nature of the business process or service. We will discuss the unit in more detail later but, until then, remember that processing one unit at a time can bring important benefits to both the processor and the customer.

This is fairly intuitive in services; clearly, customers are not happy when service providers batch, or attempt to process, more than one customer at a time. The quality of the service almost always deteriorates and, at the very least, the customer perceives a lack of attention. Batch processing also leads to increased lead times and processing times, irritating the customer, and further burdening the service provider. Finally, well-intended service providers create non-value-added processing waste for themselves as they bounce from one customer to the next, whether in food services, call centers, health care, or other service environments.

And what happens when you try to batch-process in an office environment? For information processing in an office environment, the quality of the process can deteriorate with batch and queue processing. Mistakes tend to affect a greater number of units, since problems are most often discovered during downstream processes. And that means that *everything* processed at or around the same time (in other words, in that batch) becomes suspect. At the very least, the organization now needs to examine the work to verify that no other units are affected. For example, many companies use Electronic Data Interchange (EDI) technology to process orders between parties (i.e., the customer and supplier). Most often, these are transmitted in batches once or twice a day. When a problem is discovered with one order, all orders in the same transmission are questionable, depending on the nature of the problem.

This is where the Lean concept of *flow* comes in. The goal of flow is to eliminate the "hurry up and stop" processing of information or service resulting from batch and queue processing. We just need to look at the total lead time for a process, which includes all queue and wait time, and compare it to the process time to determine the amount of interruption that occurs in the current process and the opportunities to minimize it. How do we reduce the amount of interruption? There are three different approaches to the application of flow in the office and services that we can take, which we will explore next.

Approaches to Flow in Office and Services

Here are our three alternatives to approaching flow in office and services:

- Combining activities into a single role or responsibility
- Different roles or functions performing activities in a continuous flow manner
- Performing activities in parallel with other related activities

Each of these approaches has its advantages and disadvantages, which we will explore in depth along with factors that you should consider to determine the appropriate approach for you.

Combining Activities

For the *combining activities* approach, we will use value stream mapping icons (see Figure 3.1). On the left of the figure, you can see three processes (1, 2, and 3) that are performed in series, each performed by different people, functions, or departments. For every hand-off, there is the opportunity for a queue to form. This is particularly the case in the multitasking environment often found in offices; the probability that a person receiving the work is ready to immediately process it decreases in a multitasking environment, where a person performs multiple activities in the course of the day, week, and month. The likelihood of a queue forming increases; lead time increases with more queues; and more work is in process, which can give rise to other issues (e.g., losing information, additional sorting).

On the right of Figure 3.1, you can see that the three processes (1, 2, and 3) have been combined into a single role and assigned to one person. That person is responsible for performing all three activities. This approach qualifies as flow in that queues cannot form between processes anymore (if properly done), because the hand-offs have been eliminated. Lead time will be reduced as a result. Now, a queue could still form prior to this. We will talk about how to deal with that when we discuss demand rates and capacity. For now, let us just say that if the demand is such that we need three people to perform processes 1, 2, and 3, then we would triplicate this role, and each person assigned to it would perform all three activities. Overall process time is often reduced as well, as different people do not have to familiarize themselves with the same work. Let us say we need to create a quote for a particular service. In batch and queue processing, each person at each hand-off often must review the same information in order to proceed with their part in creating the quote.

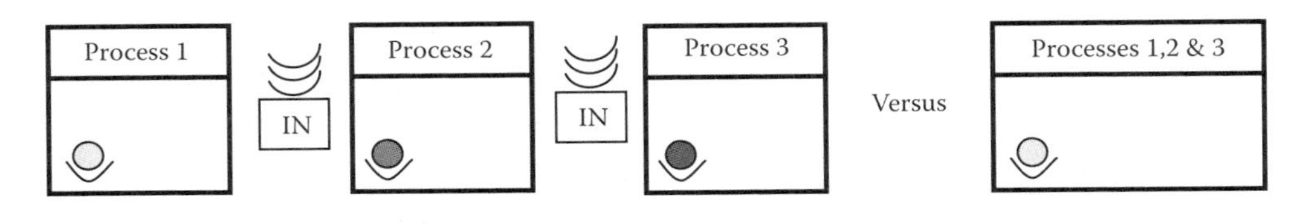

Figure 3.1 Batch and queue processing versus combining activities.

SIDEBAR Value Stream Mapping

Value Stream Mapping is a technique to visually depict the flow of materials, information, or a service. As with all mapping techniques, it makes use of symbols or icons, each of which has a specific meaning. By using icons, you can convey much information in succinct ways. The icons can identify waste and highlight issues with flow. In Figure 3.1, we used the following icons:

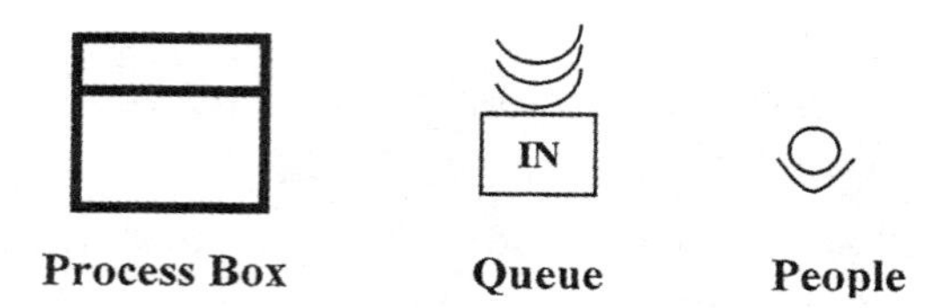

Value Stream Mapping is different from most other mapping techniques in that it also includes important process data, thereby providing a more complete understanding of the process. It has been called the "assessment and planning tool of Lean practitioners." *The Complete Lean Enterprise: Value Stream Mapping for Administrative and Office Processes* (Locher/Keyte, Productivity Press, New York, NY, 2004) is the preeminent reference on the subject.

In most service environments, customers also welcome this approach. One only has to recall the experience of calling into a technical support center—say, for a computer related problem—only to be transferred from one person to another. Customers become rightfully dissatisfied as they have to repeatedly describe the problem to a different person.

What are the possible disadvantages to this approach? The main concern is the complexity of the work. By combining activities 1, 2, and 3 and asking one person to perform them, have we created a role that is unreasonable to expect one person to adequately perform even with some amount of training? You would not be faulted for initially assuming so. After all, the work must have been divided among three people (or departments) for a reason. But we need to challenge that assumption. Often, there is no reason other than that "we have always done it that way." There may also be assumptions that only certain departments can perform particular tasks. Again, you need to challenge those assumptions.

People can often take on other responsibilities, in time, with some training and education. Of course, the existence of standard work for each activity will greatly shorten the learning curve here. In practice, people must look beyond titles and departments and focus on the activities themselves. What is really involved? If we can shift our focus from titles and departments to activities, we have overcome a major hurdle. Focusing on activities is one part of the "DNA of Lean," along with "pathways" and "linkages" (Steven J. Spear, H. Kent Bowen, "Decoding the DNA of the Toyota Production System," *Harvard Business Review*, September 1, 1999).

A pathway is the movement of the information or service from activity to activity, often involving a hand-off from one person to another. A linkage refers to the fact that the output of one activity is the input to the next activity. We always need to make certain that the output of one process is *exactly* the input that the next process needs. These activities also need to be coordinated with each other. For example, it often makes sense to coordinate the timing of activities in order to maximize flow and maintain customer satisfaction. Certainly, by combining activities, the pathways and linkages will be greatly simplified.

Continuous Flow Processing with Multiple Roles

Once again, we can use value stream mapping icons to visually depict how to use continuous flow processing with multiple roles (see Figure 3.2).

To the left of the figure, you will see the same icons you saw in Figure 3.1: three different processes, each performed by a different person or department. On the right side of the figure, there are still three different people performing the three different activities; however, they are working together, colocated, with no or minimal queues forming between them. In other words, a cross-functional, colocated team following "one piece" or near "one piece" flow is processing the information in office "cells" or "pods."

The key to this approach is not to simply bring together the different activities, but to make sure that the work is balanced or evenly distributed between each person (see Figure 3.3). Only then can you reduce or even eliminate the possibility queues forming. In other words, you need to carefully consider which activities need to be performed by whom. Think about various activity combinations. How can you organize them to achieve the best balance? Most often, that means incorporating the first approach, combining activities.

The advantages of this approach include reduced lead times, since you have eliminated your queues, and improved quality, since the supplier and

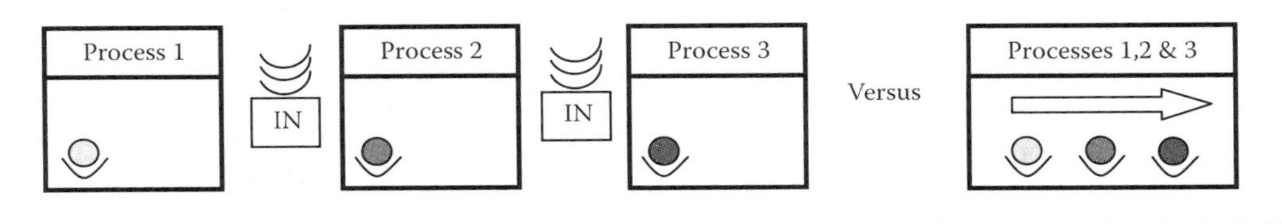

Figure 3.2 Batch and queue processing versus cross-functional, colocated teams.

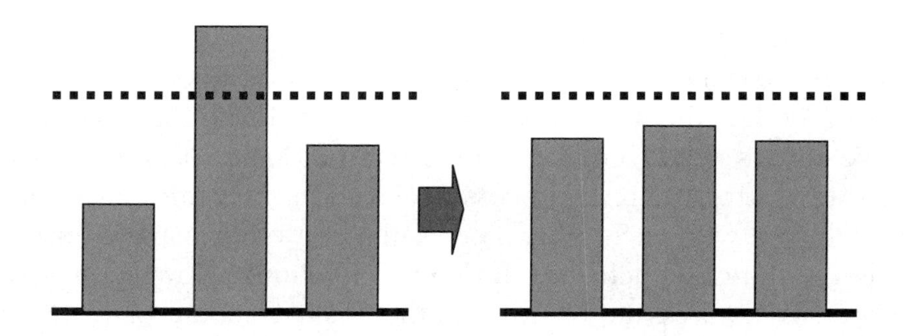

Figure 3.3 Achieving process balance.

customer can communicate directly and in a timely way about quality issues. Another potential benefit is reduced process times, most often as a result of improved quality. Process time often includes some non-value-added correction waste, which becomes a natural *part* of the process, and therefore goes undetected or at least unchallenged. The proximity of the supplier of information to the customer will almost always give rise to improved quality, as suppliers begin to better understand their customer's true needs.

Additional process time reductions can be traced to improved communication and decision making. Lengthy meetings called to ensure communication between departments or to make a decision among different functions can be replaced by 5–10 minute "huddles." Other benefits of this approach to flow include the improved ability to cross-train between functions over time (a very difficult task when different functions are located in different areas), which leads to improved flexibility and responsiveness. For example, let us say that demand for a particular role or function increases, not on the team itself but within the team. The cross-functional, colocated team approach more easily allows for other team members to lend assistance. Many alternatives exist for the physical layout of an office cell or pod. One possibility is illustrated in Figure 3.4.

What are the possible disadvantages, real or perceived, of this approach? One frequent argument against this approach involves the impact on existing functional management structures, that is, "How can I manage my people if they are located in teams in different areas of the facility?" Well, the fact is that most organizational structures (i.e., the "organization chart") have little to do with business processes or process flow. The functional or departmental nature of traditional structures actually impedes flow and contributes to non-value-added time. Lean thinkers do not get caught up in the management argument. Lean leaders go to the "gemba," or where the work is actually

Figure 3.4 An example of an office cell. Three different functions have been colocated. Position 2 is duplicated depending on the level of effort required for this task to meet demand.

performed, to observe, identify nonstandard conditions, ask questions, confirm that standard work is being followed, etc. All of this can continue to take place regardless of where people are physically located. The implementation of visual management techniques that are *always* a part of implementing flow systems in the office (e.g., office "cells," office "pods") makes process management vastly easier. So, this disadvantage does not typically hold up.

Do reporting relationships need to change as a result of implementing cross-functional, colocated teams? Ideally, yes, but it is *not* necessary. In a perfect world, the team or teams will report to a "Value Stream Manager," or what some call "End-to-End Process Managers." Imagine for a moment that your organization has a manager for the entire "Order-to-Cash" process, rather than the information flow crossing departments under the responsibility of several managers. How about a single "Requisition-to-Pay" value stream manager, rather than different managers for purchasing, material management or planning, and accounts payable? How would your organizational dynamics and politics change? Remember, as we discussed in Chapter 1, you can organize by value stream without changing reporting relationships. All you need is the support of the existing functional managers.

Another disadvantage commonly voiced is the loss of interaction within functions between people of similar knowledge and skills. For example, accounting professionals will no longer be able to conference among themselves to share experiences, knowledge, etc. This can be a real disadvantage, but only if left unaddressed. A simple fix is to arrange a periodic meeting of accounting personnel within the organization to ensure knowledge sharing.

Another very real issue involves activities that require very specialized skills and experience, and/or that support different value streams or information processes. Let us take a look at a real-life example. At an "engineer-to-order" company—where every order required some amount of design work, "contracts management" was such a function organized by customer. Contracts managers worked closely with specific customers. They possessed a deep understanding of the customer and customer needs with regard to contract negotiations, terms, administration, and ongoing relationship management. Several "quote-to-ship" teams were implemented based on order or project complexity (i.e., 1-month projects versus 24-month projects). This organization was faced with a real issue in assigning a contracts manager to one particular quote-to-ship team, because specific customers may have had projects of different complexity running concurrently (being processed by different teams). What can be done in this situation?

The answer came from the contracts managers themselves. The managers suggested that they locate themselves with a particular team at a particular time based on need. If there was a project at a certain point in time that required their involvement, they would physically locate themselves with that team for whatever time was required (see Figure 3.4). The contract managers did not worry themselves about the "nomadic" nature of this arrangement, and the results were impressive. Quote turnaround time was reduced by 75%, and order turnaround time was reduced by 50%.

Concurrent Processing

Now that we have examined your first two alternatives to flow, let us take a look at completing activities concurrently, or in parallel. Concurrent or parallel processing is another approach to flow that can further reduce lead time.

There are important advantages and disadvantages to this approach. First, you need to think long and hard about whether it is even possible in your particular situation. Many information and service processes cannot be conducted in parallel. Second, assuming that it is possible for your information or service process, you can expect managing the process to typically become *more* difficult. The queue at the convergence point after the three processes can still grow significantly if one process lags behind the others (see Figure 3.5).

This approach requires more adept process management skills to keep the convergence points balanced. The visual management techniques

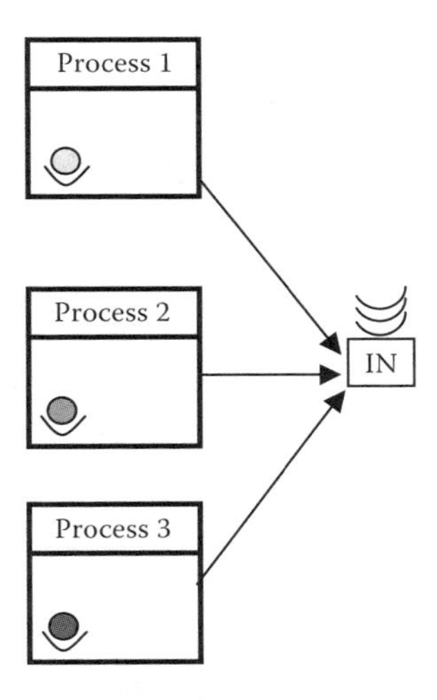

Figure 3.5 Performing activities concurrently or in parallel.

espoused by Lean (and covered in Chapter 5) can greatly help here. In Chapter 5, I will show you how to easily and quickly identify imbalances between flows and take corrective action to bring the system back in balance to maximize flow.

I found a service example of concurrency in my personal experience during a visit to a root canal specialist. As a patient, I was immediately led to the examination room upon arrival. Various administrative activities were performed in parallel to the doctor's examination. In most situations, the administrative tasks would have been completed while the patient was still in a waiting area. The concurrent approach removed an estimated 15 minutes from the overall lead time of the office visit.

Concurrent processing has provided breakthrough results in some information-intensive processes, such as product development. So, while it may be more difficult to manage, do not overlook the benefits.

Designing Flow Systems for Office and Services

There are six basic steps to designing flow systems for information and service processes:

1. Identify the activities involved.
2. Determine the demand rate for each activity.
3. Determine the resource requirements.

4. Identify roles and responsibilities, including standard work for each.
5. Determine training and cross-training needs.
6. Develop visual management techniques that will be used to manage the system over time.

Identify Activities

First, you need to specifically identify the activities to be performed in your envisioned flow system. Again, clearly defining activities is a part of the "DNA of Lean"; it is a critical step and can lead to breakthrough results. Further, this step requires people to look beyond their narrow definitions of functional roles. What activities will be performed as part of the flow system? Remember, there may be some activities that just do not fit within our three potential flow systems; however, you do need to challenge any resistance to including particular activities, and you can do so by referring back to effective responses to the more common arguments. Once you have completed this important step, your next focus is the "demand rate" for each activity.

Determine the Demand Rate

It is important to understand the demand rate, or "Takt Time," for each activity to be performed in the flow system. Takt Time is one of the most misunderstood concepts in Lean, and people have struggled mightily with its application to office and services. First, we must understand its purpose.

SIDEBAR

Takt is a German word that means interval of time. It was first used in manufacturing to set a pace or tempo to the production of airplanes in the 1930s.

Takt Time is simply the expression of demand in terms of time—minutes per order, hours per quote, etc. Its inverse is a more commonly recognized measure—orders per hour, quotes per day, etc. In both cases, it is a measure of demand, used to determine the required capacity to meet demand. This can be a breakthrough concept for office and services, in that many organizations have not sufficiently performed capacity planning analyses to ensure that capacity is in line with demand. Now, some highly transactional service environments (e.g., call centers) have already figured this out, but the concept is still new for most others.

The equation for Takt Time is quite simple:

$$\text{Takt Time} = \frac{\text{Effective Working Time}}{\text{Demand}}$$

So what is the difficulty? Several issues can arise. First, let us look at the numerator, "effective working time"? How do you decide what figure to use? Simply put, it depends on the scheduled hours for the person expected to complete the activity. For the first round of analysis, we will assume a number of shifts in the day and a full day of effective working time. This is an easy calculation to make, since most organizations know the total number of hours they expect to schedule either the office or service delivery; in other words, the hours the organization or department is "open for business." This is most often driven by the external customer, though a subsequent analysis of capacity may make an organization reconsider its scheduled hours. You may need to add more hours to the business day to complete all required activities to meet demand and maintain customer satisfaction. For example, the office of an insurance broker may need to be open longer in order to provide customers the ability to drop in before and after their business day is over, and to allow adequate time to complete all required paperwork processing. Of course, any planned downtime; for example, lunch breaks or shift start-up meetings, must be removed from effective working time.

How about the denominator, demand? What should we use here? First, our denominator must be compatible with the numerator in terms of the unit of measure. If effective working time is expressed in daily terms, then demand must also be expressed in daily terms. If effective working time is expressed in weekly terms, then demand must also be expressed in weekly terms. Next, we need to determine the appropriate measure of demand. This trips a lot of people up. What is an appropriate measure of demand? It may be different for different activities, but it needs to relate to the work, and more specifically the work content, being performed.

For order processing, expressing demand in terms of orders may not be appropriate since the process time can vary greatly from order to order. A better expression of demand may be line items, since line items directly relate to order process time. However, the time required to process an invoice may not depend on the number of line items; in this instance, the number of orders may be sufficient.

The discussion can get even more interesting in highly variable, project-oriented work such as that found in "design-to-order" companies, information technology (e.g., software development), and similar situations. How can you express demand if the time to complete a project can vary from 10 to 1,000 hours? In designing a flow system for this kind of workplace, your best bet is to use historic information depicting the actual hours used in past projects, assuming that it is indicative of future demand, which is often the case. Otherwise, the historic information will need to be adjusted either up or down, based on anticipated future demand.

Of course, demand can vary, so you may need to complete several calculations of Takt Time. When designing the flow system, it is often worthwhile to calculate Takt Times that correspond to busy periods, slow periods, and average periods. This is particularly the case for seasonal businesses that experience *sustained* periods of changes in demand. Short-term changes in demand can be handled through other means (i.e., with pull systems, which we will cover in Chapter 4) and are often not considered at this point of the process unless absolutely necessary.

Next, let us take a look at how we can use Takt Times to determine resource needs.

Determine Resource Requirements

To determine your resource or people requirements, you need to compare your various Takt Times to the process times for the corresponding activities. By dividing process time by Takt Time, you can determine the number of resources required (see Table 3.1).

Before you can do this analysis, you need to determine the process times for each activity, so you will need to conduct some form of time study. And again, you will need to repeat the analysis for different Takt Times, if demand is expected to vary significantly at particular periods of time. This analysis will give you an idea of the number of people to be involved in the flow system. In the calculation in Table 3.1, a total of 3.85 people are required, so let us say that we need four people.

How can you determine your resource requirements in the highly variable project-oriented environment previously mentioned? For example, let us say we currently have three projects to be completed, which vary significantly in terms of complexity. Project "A" requires 40 hours of process time, and Project "B" requires 10 hours. A third project, Project "C" requires approximately 30 hours to complete (see Figure 3.6).

Table 3.1 Determining Resource Needs

Activity	*Process Time*	*Takt Time*	*# of People Required*	*Comments*
Enter Order	5 min per line item	10 min per line item	0.5	
Send Acknowledgments	2 min per order	20 min per order	0.1	
Design Order	60 min per order	20 min per order	3.0	
Invoice	5 min per order	20 min per order	0.25	
		Total	***3.85***	

Project	Estimated hours
A	40
B	10
C	30
Total	80

Figure 3.6 Determining resource needs in a project environment.

Is our current number of projects, three in the example depicted in Figure 3.6, a good estimate of future demand? Probably not, since the amount of time required to complete each varies so much. The number of hours of work required to be completed during a period of time will be a better expression of demand on the resources expected to complete the projects. In the present example, we have 80 hours of work to complete. Let us further assume that, because of customer dictates, we need to complete this work within one 40-hour work week. So, what kinds of resources will we need to complete the work? We need two people to meet the demand in the 1-week timeframe. This calculation is displayed in the following equation.

$$\begin{aligned}\text{Number of required resources} &= 80 \text{ hours of work} \div 40 \text{ hours of work time}\\ &= 2 \text{ people}\end{aligned}$$

Once again, we can arrive to a conclusion regarding resource needs by calculating Takt Time and dividing it into process time. It is a simple but very effective calculation.

Let us explore one more resource requirement scenario. What do you do if there is no customer-dictated level of service? Many information processes performed in an office environment do *not* add value to the external customer but, instead, are performed to meet the needs of internal customers—fellow employees of the company. The level of service to internal customers for a particular process is not always well defined, and implementing Lean provides an opportunity to add clarity. With some prompting, the internal customer himself can often provide this clarity ("I really need it by the first of the month"); alternatively, the service provider can set the expectation in the absence of a true customer need ("I will get it to you within a week"). Either way, by establishing such "service levels," we can assess the demand in a period of time to determine the resources needed to meet that demand.

Identify Roles and Responsibilities, including Standard Work

Now let us look at Table 3.1 again, a little more carefully. In our example, three people will be required to "design orders," and a total of 0.85 people will be required to perform the other three activities listed. At this point, we should start to consider *combining* the three activities "Enter Order," "Send Acknowledgments," and "Invoice" into one role. In other words, it is time to challenge existing roles and preconceptions of *who* can do what activities.

In our example, the required resources for each role are matched to their respective demand rates or Takt Times. This matching will result in a balance of workload between the resources, each at or near the respective demand rates, and sufficient capacity for each role to meet the demands for all of the activities each is expected to perform. Further, there will *not* be excess capacity in any role that can create imbalances in flow and allow for queues to form between roles.

In a perfect world, we could load each role at exactly the estimated Takt Time, but it does not work that way in practice. In the example in Table 3.2, a total number of 3.0 people performing the "Design Order" activity are required to meet demand—but that is only if everything goes according to plan.

To prepare for variability in demand, variation in complexity of the work, or additional, intermittent task work, we need to build in a modest amount

Table 3.2 Determining Resource Needs

Activity	Process Time	Takt Time	# of People Required	Comments
Enter Order	5 minutes per line item	10 minutes per line item	0.5	
Send Acknowledgments	2 minutes per order	20 minutes per order	0.1	
Design Order	60 minutes per order	20 minutes per order	3.0	
Invoice	5 minutes per order	20 minutes per order	0.25	
		Total	***3.85***	

of excess capacity. Typically, 10%–15% excess capacity will do the job. In our example, the total of 0.85 people required to perform "Enter Order," "Send Acknowledgments," and "Invoice" activities is rounded up to one person. This exemplifies what is done more commonly in practice and gives us a 15% excess capacity cushion for variability in the work.

SIDEBAR

Sometimes we need to look at an activity more closely to determine if there is variation of significance within it that needs to be addressed. For example, perhaps the time to enter an order significantly varies based on how it is received (telephone, fax). In these cases, you need to determine a Takt Time for each type of order. Calculate the Takt Time for telephone orders and compare it to the current time to process telephone orders, and do the same for fax orders. The calculations to determine the number of people required are the same as those displayed in Table 3.1.

In our example, we have calculated that we need a team of four people to perform our order processing tasks. Next, this team needs to have a meaningful discussion to identify and define their roles and responsibilities. What really can or cannot be combined? What other activities are we expected to perform? For example, is this team expected to provide order status to customers? If so, how is this activity going to be integrated with general order processing? Once these preliminary questions have been answered, the team can move on to defining standard work for each role, in order to maximize flow.

As you will recall from Chapter 2, the elements of standard work are defined as the activities to be performed in a desired sequence, the time and timing for performing the activities, and the "key points," which includes "how" and "why" the work is to be performed in a particular way. Documenting standard work in simple ways will greatly help as we move

Process: Mortgage Processing Takt Time: 4 minutes Operation: Application Processor			
Activity	**Key Point(s)**	**Time**	**Time Scale (mins)** 0 1 2 3 4 5 6
1. Check application for completeness	Do not begin if incomplete	1 min	
2. Enter into system	Must input all required fields	2 min	
3. Create application folder	Must have application number	0.5 min	
4. Pass to credit check	Place in proper timed bin	0.25 min	

Figure 3.7 Standard work combination form, including Takt Time.

into Step 5. I provided several examples in Chapter 2. For another possible example and format, see Figure 3.7. This format includes the Takt Time that must be met in order to meet demand. Each activity and its respective process time is listed in the desired sequence. Note that, when added together, the process times for this particular operation are just under the Takt Time. This format is referred to as a "Standard Work Combination Sheet" and is commonly used in highly repetitive work environments where the same activities are frequently and repeatedly performed.

Before moving onto the next step, let us discuss one more topic. Most office "teams," "cells," or "pods" will involve multiple processes, activities, or tasks. A "plan for every process" (refer back to Chapter 1 for an example) will be defined and followed by team members so that flow through the team is maximized. Team members cannot revert to large batch processing of any particular task if it results in impeding the flow of the other tasks. To prevent this from happening, particular tasks must be performed at appropriate frequencies, which should be defined in the "plan for every process" as a part of the standard work developed for each team member.

Determine Training and Cross-Training Needs

Once roles, responsibilities, and standard work have been defined, we need to look at team members' existing skills. Any "gaps" in knowledge, skills, and abilities between the roles to be performed and the people who are expected to perform them must be defined. Commonly, some amount of training or cross-training will be necessary. With defined standard work, this will be much easier, and again, by using effective training techniques (Job Instruction [JI]), the learning curve will be greatly reduced. Once the required training has been completed, it is time to implement your envisioned flow system.

Develop Visual Management Techniques

Your next step is to implement effective visual management techniques. You need visual management to sustain the flow system and improve it over time, which is the real objective of Lean. We will cover visual management techniques in detail in Chapter 5. For now, you should be aware of an important concept relating to Takt Time.

Although the primary purpose of Takt Time is simply to identify the time needed to conduct a process to comply with demand, another purpose of Takt Time is to set a "pace" or a "tempo" to the processing of work. This is not typical of many office and service environments; instead, people will process work "when they get to it," or until "somebody screams for it." This often leads to work going undone until the last minute. Even if the customer does not require a specific response time, setting service levels allows us to establish a pace for the work. Importantly, progress can then be monitored to identify situations in which work is not being attended to, and visual management techniques can be implemented to identify these situations in a simple and timely way. This concept of monitoring whether demand is being met in a simple and visual way has been called "Takt Image" or "Takt Awareness."

One such method of providing Takt Image is shown in Figure 3.8.

A board has been set up to track the output of a cross-functional team every 2 hours. Some refer to this as a "Plan versus Actual" board, while others call it a "Pitch" board. The team's expected output, marked in 2-hour intervals, is identified in the column marked "Plan." The team tracks their actual output in the column marked "Actual." The "Comments" column is meant for notes to add clarification to the figures. "Pitch" refers to the frequency with which performance is reviewed; in this example, the "pitch" is every 2 hours.

Hours	Plan	Actual	+ / -	Comments
8:00 – 10:00	10	5	- 5	Computer down for 20 minutes
10:00 – 12:00	20	15	- 5	
12:30 – 2:30	30	28	-2	Able to catch up somewhat
2:30 – 4:30	40	38	-2	

Figure 3.8 Providing Takt image.

Pitch boards should be reviewed periodically and recurring problems identified. For the example provided in Figure 3.8, let us say that computer downtime often resulted in a lower output than planned. To improve performance, the computer issue needs to be addressed by the appropriate individuals. The board then becomes part of an overall visual management system, which we will cover in depth in Chapter 5.

Typical Results

In my experience, typical results and benefits of the application of flow concepts to office and services have included

- Lead time reductions of 50% to 90%
- Process time reductions of 20% to 40%
- Quality improvements (e.g., "defect" reduction) of 25% to 75%
- Increased employee satisfaction with the workplace environment

These are impressive results, and you would be forgiven for being skeptical. However, they are real results; for lead time, for example, most queue times will be reduced and, typically, up to 95% of lead time is associated with queues. Applying flow concepts will directly address, and minimize or eliminate, these queues.

Remember our discussion of the basis for the process time reductions? These included the reduction of correction waste often embedded in process time; the reduction in non-value-added processing waste associated with interruptions; sorting and resorting work, refamiliarizing oneself with work, elimination of redundant activities, and more effective communication and decision-making processes. Altogether, this can add up to a 40% reduction in process time. And do not forget the benefits to quality as a result of the application of flow concepts, including improved awareness of internal "customer" needs and a more timely identification of quality issues, to name just a few.

So, let us focus on the last, but by no means least, benefit: associates will find the resulting work environment to be more satisfying. It is true that this will not be the case in the beginning. Experience has shown that 3 months is typically required for people to "settle in" to the new process (and sometimes longer). But, after sufficient time, most employees speak very positively about the new process. Many causes of historic frustration are directly addressed by the application of flow concepts. First, work is "balanced" between resources, thereby addressing the concerns of some individuals that they are "doing twice as much as others." There is an element of perceived fairness that comes with process balancing.

Second, you will find that your employees have a greater sense of achievement. While there may be initial resistance to assuming different duties and cross training, most people will come to welcome this over time. Most humans do not want to perform mind-numbing repetitive work. The combination of activities, as well as the flexibility encouraged by cross-functional, colocated teams provides greater opportunity to learn and perform new duties.

Third, there is a greater sense of "team," and most people, though not all, welcome this. Humans are generally social beings who need a sense of belonging, and that can be provided by cross-functional, colocated teams if that is the approach you take. Finally, much frustration and stress can be removed with the improved organization that will be achieved.

Summary

In your office and service environment, as in manufacturing environments, one of the goals of Lean is to achieve balance to maximize flow through a value stream. Although the concepts we have covered in this chapter can greatly help you move toward this important goal, in practice, value streams can still become imbalanced from time to time. This can happen despite all of your analysis, planning, and best intentions. Variation, although greatly reduced, will still exist in office and service processes even after efforts to standardize work and to achieve balance have been put into place. But not to worry, because Lean provides another concept to address temporary imbalances in the flow of information and services: "pull," which we will cover in the next chapter.

Chapter 4

Creating Level Pull in the Office

In this chapter, we will cover a key Lean concept: pull. As a general rule, we always want to apply the flow concepts covered in Chapter 3 first, and then apply pull where necessary. But there are exceptions, and often, pull can be easier to implement than flow, as it represents *less* substantial change. For example, an organization can implement pull without the need to physically reorganize the office layout. Smaller changes are viewed as easier medicine to swallow, and a nice early step in the implementation of a Lean office or service organization. Further, as you will see later in this chapter, the application of pull can trigger other important practices such as cross-training. This will lay a foundation for the implementation of the flow concepts covered in the Chapter 3.

Pull, and its application to office and service processes, continues to greatly confuse people. Pull is, simply put, a method of controlling the flow of resources based on actual demand or consumption. In its most basic sense, it is a decision-making tool. In manufacturing, for example, it helps organizations decide what to make and when to make it, and what to buy and when to buy it. In offices and services, the resources to be controlled are information and people. Again, despite the differences among these organizations, the *same* pull concepts used in manufacturing can be applied to office and service organizations. They can help office and service workers decide what to work on and when to work on it, thereby maintaining customer service while preventing overproduction.

SIDEBAR Overproduction in Office and Service

The concept of "overproduction" can be difficult to grasp when considering office and service processes. Overproduction in offices and services can be defined as *more* information or service than is needed, *sooner* than is needed by the downstream or customer process. The "sooner," or timing element, is often overlooked. While "more" is not always a concern in an office or service environment, "sooner" most certainly is. For example, training can be considered a service. If training is provided too early and the trainee has no opportunity to apply the new knowledge or skill, the value of the training declines. The trainee will lose much of what was learned over time in the absence of the opportunity to apply it.

Forms of Pull Systems

The *Lean Lexicon* (Lean Enterprise Institute, Fourth Edition, March 2008) describes two forms of pull systems—supermarkets and sequential pull. These two systems can also be implemented in combination:

- Supermarket pull systems—In this type of pull system, a process will have a queue or storage point located after it that holds a particular amount of each output (e.g., product, information) that the process can produce. The process simply replenishes what has been consumed by the downstream or customer process per a given set of decision rules (e.g., order points, order quantities).
- Sequential pull systems—In this type of pull system, a process has a queue or storage point (located after it), but not necessarily of each output (e.g., product or information) that the process can produce. What is actually in the queue can vary at any time. Essentially, it is a "process-to-order" system. The process replenishes based on the status of the queue at any time per a given set of decision rules (e.g., order points, sequencing rules).

Each type of pull system is depicted in Figure 4.1 using Value Stream Mapping icons. On the left, you will see a Supermarket pull system. The shelf-like icon between Processes 1 and 2 is the supermarket icon. The figure on the right is a sequential system. The arrow-like icon between Processes 1 and 2 is the sequential pull icon. In this example, the desired sequence is "first-in-first-out," or "FIFO," which is noted in the arrow. As you read the preceding descriptions, you can see that the nature of work found in an office environment and in services naturally lends itself more

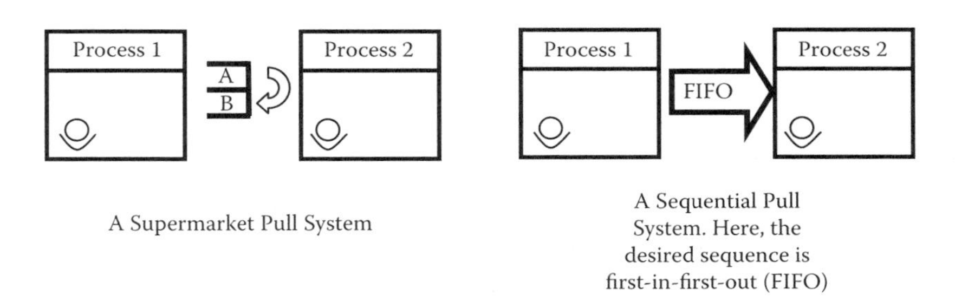

Figure 4.1 Depicting two types of pull systems using value stream mapping icons.

to sequential, rather than supermarket pull systems. After all, it is not practical—and often not even possible—to have a given amount of *every* type of information that an office process can produce in a queue (i.e., on a "shelf") at all times. Nevertheless, there are some Supermarket applications that you can use in your office environments.

The most common Supermarket pull system in office environments is for office supplies, since it is easy to establish reorder points and order quantities for these items, and little effort is then needed to manage the inventory of office supplies over time.

Let us take a look at an example of an interesting application of an electronic Supermarket pull system, for the auditing process in a publicly traded company that I worked with. This organization determined that many types of information were frequently and periodically requested by public accounting firms as part of their audit processes, including evidence or examples of particular transactions, approvals, and other types of information. In the past, the organization's auditors would request this information from the company's accounting personnel, creating frequent disruptions as personnel stopped whatever they were doing to respond.

To minimize these disruptions, the publicly traded firm decided to create an electronic supermarket. Each month, accounting personnel generated examples of assorted types of information that they would store electronically. The auditors "pulled" the information that they needed, but without creating disruptions in the accounting department. The company replenished the supermarket monthly. This still represented some amount of overproduction, which is inherent in all Supermarket pull systems. However, the organization estimated that approximately 85% of all of the disruptions associated with the audit process were eliminated, while maintaining customer (i.e., the auditors) service. This system is really no different from systems within a grocery store, where milk is replaced when it has passed its expiration date.

In general, sequential pull systems are most frequently applied within an office or service environment. Using a sequential pull system, the office or service organization places a limit on the queue, which, when reached, triggers a decision. In the case of a maximum limit, it could mean that the supplying process (i.e., the "supplier" of the information) decides to stop processing a particular type of information and begins processing a *different* type of information. There is no need to process more information or overproduce beyond the identified maximum, since the downstream or customer process cannot process it in a timely manner. In this case, the supplier might as well work on something else.

In the case of a minimum limit, this would signal the supplier to return to processing a particular type of information, since the downstream process will be ready for it. You can easily see that the flow of information *and* the "supplier" can be effectively controlled by establishing such basic decision-making rules. Further, with this pull system in place, the supplier can self-manage, requiring little or no direction from a supervisor or manager.

Even in a sequential pull system, you will have overproduction, but you are also controlling the queue. Nevertheless, there is an opportunity for the customer, or downstream process to select whatever work it wants from the queue. Because of this, you need to establish clear rules throughout the value stream to ensure that people are working to consistent priorities. Therefore, you need to enhance your decision rules by defining the desired sequence. The most common desired sequence in office and service environments is "first-in-first-out," or FIFO, as shown in Figure 4.1. However, depending on the type of work conducted, there may be other desired sequences; for example, in a project environment, the desired sequence is most often by due date. There is no need to work on projects on a FIFO basis if one project is not due for quite some time relative to other projects. Depending on the application, organizations may choose other desired sequences; for example, by complexity. Figure 4.2 illustrates a combination of sequencing rules. First, the projects are initiated at Process 1 by due date. Thereafter, they are processed on a FIFO basis. In other words, Process 2 will simply process the projects in the order that they are received (i.e., first-in-first-out). This will maintain the desired priority (due date) that was already established at Process 1, keeping the scheduling of work simple for Process 2 and all subsequent processes (not shown). No need for separate schedules for each process, or complex computer-based scheduling systems.

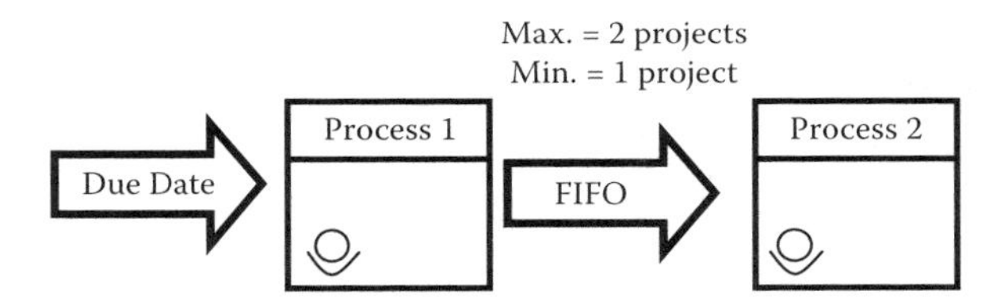

Figure 4.2 Sequential pull system.

At this point, you will recognize that all pull systems share some common characteristics, including

- Visibility of a queue of work
- Limits defined for a queue
- Defined rules for the queue when the limits are met
- Use of visual signals (i.e., kanbans) that are worker managed

We will discuss each of these individually next.

Visibility of Queues

As we have already seen, one key aspect of implementing Lean is to make the queue of work visible. These days most information resides in an electronic form, such as your e-mail inbox. This has actually made it *more* difficult to make work visible than in the past, when most information took the form of hard copies. Often, organizations need to develop other electronic tools in order to make the queue visible.

Another challenge, again, is behavioral. People can be reluctant to visibly display their pending queues of work for others to see. Why would this be the case? Perhaps people are concerned with the response that they will receive, or how they will be perceived by their co-workers or manager. Often, simple worker-managed means that can be used to make the queue of work visible are ineffective or ignored altogether, because people are reluctant to use them.

An example of a simple visual method for displaying work in progress is illustrated in Figure 4.3. A "Rolodex" of numbers is located at the drop-off

Figure 4.3 Providing visibility to the queue of work.

area or "inbox" for a particular operation. As the supplier drops off more work, he or she flips to the correct number to reflect the amount of work now in queue and in-process at the next operation. It is that easy. If the supplier drops off two orders, he or she adds two to the existing figure and flips to the correct numbered card. As the processor completes an order, he or she subtracts from the figure and flips to the correct numbered card as the order is passed on to the next operation. When you use such simple methods, it is easy to maintain visibility of the amount of work in-process and in queue at all times.

Establishing Limits on Queues

But how do you determine the acceptable limits for the queue of work? When sizing your pull system, you need to consider the following:

- Required customer service levels—the overall goals established for meeting the needs of the customer (i.e., lead time)
- Organizational goals for completing particular processes or activities

For a series of processes, say, an entire value stream, the total amount of work in process—in queue or actively being worked on—at any time must be limited to ensure that lead time goals are consistently met. When more

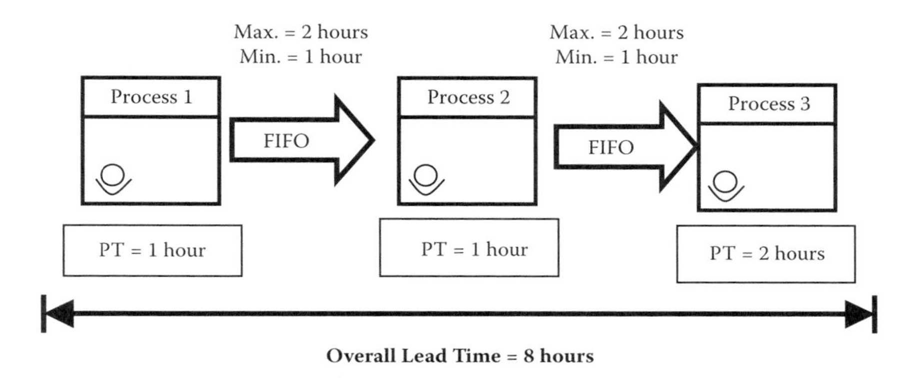

Figure 4.4 Implementing a pull system through a value stream.

work is in process, you will have a longer lead time through the value stream. Further, if the work-in-process is allowed to significantly vary, the lead time through the value stream will substantially vary as well, creating a *very* unpredictable system. You will need to set limits for individual processes in such a way that you control the total overall work-in-process, thereby maintaining consistent and acceptable lead times.

Typically, what is more important than the *amount* of work in the queue is the time equivalent of the queue—the time that information, a customer, or a project has been in the queue. For example, if a call center has established a service level in which no customer will have to wait more than 45 seconds, then the queue must be monitored to identify when callers are waiting longer than the identified amount of time. It is not the number of callers in the queue but the length of time in the queue that is important.

Pull systems are often observed in retail operations, though they are not always formally executed. In a check-out area, additional registers will be opened when the lines grow to be too long and customers are waiting more than a reasonable amount of time. Additional resources are "pulled" from other areas to work the registers until the wait time decreases to an acceptable level, often at the judgment of the store or shift manager. What is often missing is a specific limit placed on the queue, which could easily be established, to trigger this decision. Figure 4.4 uses value stream mapping icons to illustrate the real objective of limiting queues in an overall value stream as part of a pull system implementation. Let us say that the goal is to move an order through all three processes within an 8-hour period (i.e., the overall lead time). If the total process time (PT) is 4 hours (1 + 1 + 2 = 4), then the total queue time through the entire value stream cannot grow greater than 4 hours, or the 8-hour goal will not be achievable. In this example, we determined that we needed to limit the queues prior to Process 2 and Process 3 to 2 hours each, which will ensure that the overall lead time will be met.

One positive aspect of the electronic form that information often takes these days is that it tends to make it easier to determine time spent in the queue. In our previous call center example, telephone systems can easily track the queue time for each call. The key is to make this information visible to the people who must make appropriate decisions (perhaps to reallocate or "pull" additional resources) to maintain the service level. You can use similar techniques to provide visibility on other types of information, say, orders that are received via the Internet, by e-mail, or on fax servers. And certainly, you can develop a means to monitor the queue of orders in order to determine if this goal is being met over time, even in near-real time.

Short of having a clock for each item in the queue, you can still establish quantity limits that have some relation to time. For example, let us say that a person in an order processing role can move eight orders per day through a particular operation. In addition, let us assume that orders need to be processed through that operation within one day. Therefore, the total in queue and in process must be limited to eight, or the actual lead time will exceed the desired lead time.

The key is to identify the proper unit of measure for the queue. Think back to the demand rate discussion in Chapter 3, where we talked about the need to identify a unit of measure of demand that relates to the work content. For example, instead of orders, perhaps the unit of measure should be line items. In any case, establishing the limits and making them visible must be part of the pull system.

Establishing Decision Rules for the Queue

We discussed several examples of decision "rules" earlier, such as reallocating resources to meet established goals or service levels; ensuring that the customer process performs work in a desirable sequence (e.g., FIFO); determining when to process information based on the status of various queues. In all cases, you need to visibly post the decision rules (see Figure 4.5).

In most cases, a key consideration in determining the appropriate decision rules for a queue is the amount of flexibility that can be instituted within the work environment and within the process. To maximize the

Figure 4.5 Decision rules posted by queue.

benefits and minimize the queue, you need a degree of flexibility. You can ensure flexibility by defining standard work for the processes involved and cross-training people in the established standard work.

Unfortunately, few organizations have adequately defined standard work for all key processes. Few organizations have sufficiently cross-trained their staff. Often, too few people within the office or service environment can perform particular tasks. This situation limits flexibility, and in turn limits the decision rules that can be established. Defining Standard Work, a foundational concept of Lean Thinking, and implementing cross-training are critical to establishing a pull system in an office or service environment.

In a Sales example, a particular company previously had what could be described as a "sales funnel," in which prequalified sales opportunities were placed in a queue for follow-up at a later time. These opportunities were identified by geographical region, creating, in reality, multiple "sales funnels." As part of their Lean office effort, the organization made these queues visible by means of a report from their Sales Contact system. They also established goals for the lead time in which they wanted to follow up on these sales opportunities. Once these queues were made visible, it became apparent that particular regions were not getting to their opportunities in a timely manner.

The company worked to develop standard work for the sales function, by region, as there were substantive differences between regions. They also changed the sales performance measurement system, making it team-, rather than individual-, based. Sales management would monitor the regional queues and redirect particular opportunities that were approaching the established lead time goal to sales associates who had available capacity. In other words, sales resources were "pulled" from one region to meet demand in another region. This was a major contributor to the 10% one-year increase in revenue at the company.

Using Visual Signals That Are Worker Managed

Whether the information is electronic or in the form of hardcopy paperwork, the signal to trigger the desired decision making should also be visible. "Kanban" is a Japanese term meaning "sign," or more generally, "visual signal." The kanban can take many forms. Remember our call center example? In that instance, the signal was a light that was tied into the telephone system (see Figure 4.6).

The light would illuminate when callers were in the queue beyond the desired "service levels." Designated personnel who were performing other duties would then answer the phones until the light was no longer illuminated. At that time, they could return to their original duties. The light triggered a "pull" of resources when the demand of calls exceeded capacity, and vice versa.

At a financial services company that processed lease applications, the information was a hardcopy application placed in a folder. A series of baskets hung in visible locations (see Figure 4.7) in the office provided queue visibility.

Figure 4.6 A light as a Kanban in a call center.

Figure 4.7 Timed baskets used in a pull system.

Separate baskets were set up for different periods of time: one for 8:00 to 10:00 A.M., another for 10:00 to 12:00 P.M., and so forth. Using these baskets, anyone would be able to identify folders that were not being processed in the timeframe established. This was an indication that one segment of the lease application process was having difficulty keeping up with demand at that particular time. Over time, associates in the office learned to respond to the visual signals provided by the folders in the baskets with little or no direction from the office manager. In other words, the pull system became "worker managed."

Worker-managed signals in the form of flags have been effectively used in design-to-order companies where customer orders must go through various engineering and other functional activities to be processed. In one example, associates were encouraged to display a flag above their cubicles indicating their progress on a particular project with regard to established processing goals (see Figure 4.8).

Figure 4.8 Worker managed flag serves as pull signal.

A green flag indicated that the associate was confident about completing his or her part of the process in the designated time. A yellow flag indicated that the associate had concerns about meeting the designated time. A red flag indicated that the associate would not be able to meet the designated time. When a yellow or red flag was posted, a decision needed to be made in order to maintain the flow of projects; most often, this involved providing assistance to the person in some predetermined ways.

In the beginning, associates were reluctant to display anything but a green flag even though problems were occurring. With significant encouragement from management, the associates realized that the flags triggered a positive response for both the individual and the organization. Over time, the system became more and more worker managed.

Leveling the System

The Lean concept of "heijunka" is a Japanese term that roughly means "levelization." Heijunka is defined as leveling the type and amount of processing over a fixed period of time. This enables processing to efficiently

Figure 4.9 Leveling output in 2-hour increments.

meet customer demands while avoiding undesirable batching and results in minimum work-in-process and lead time throughout the value stream.

Most of what we have discussed so far will result in leveling the *amount* of work in the system at any time. In Chapter 2, I provided you with methods to level the *type* of work being processed in a multitasking environment and defined an appropriate frequency or timing for each task so that batching is minimized and customer needs can still be met.

We can take the concept of leveling further still in some situations. Beyond leveling the work-in-process as previously described, you also will want to level the output of a process in order to achieve more consistent flow. It may not be enough to say that an order must be completed within a lead time of 8 hours; you may also need to specify that an order should be processed each hour. In other words, at the end of a process, a completed order should be exiting each hour. This has the affect of leveling the quantity of work throughout the 8-hour period, and is often a vast improvement over multiple orders exiting toward the end of the designated period. Although all of the orders were processed within the required 8 hours, they were not processed in a smooth and steady way.

Further, you can achieve still greater leveling as part of the pull system design (see Figure 4.9).

In this example, the total required output of invoices for an 8-hour day was broken into 2-hour increments. If the total is 20 invoices during 8 hours, then 5 should be completed every 2 hours. A similar approach can be set up for other increments of time; in this instance, the company chose 2 hours because it was a reasonable timeframe for reviewing the system and making adjustments as necessary. The 2-hour review frequency was tied to a desired "management timeframe," which we will discuss in detail in Chapter 8.

As you can see in Figure 4.9, each invoice processor is assigned a set of four bins. Each morning, invoices to be processed were placed in the bins, five to each bin. This provided the invoice processors with specific, 2-hour goals and allowed them to level their output throughout the day. The manager of the department reviewed their progress every 2 hours. If a bin still contained an invoice or invoices after the designated time, it was an indication that the processor was not able to keep up with demand. Again, this serves as a pull signal, or kanban. The manager could assign the other processor to help, or the manager herself could actually process an invoice or invoices. In either case, the lead time goal for invoice processing continues to be met, and the leveling of the output of invoices is maintained.

Steps to Implement Pull Systems

Now that you understand the elements of all pull systems, let us review the steps to implement pull. The typical steps for implementing pull are

1. Identify the locations where queues are expected to form
2. For each queue, identify the means to provide visibility
3. Establish limits for the queue
4. Define the rules for the queue when the limits are met, as well as for the desired sequence of processing
5. Train people in the pull system and initiate the system
6. Monitor the system for effectiveness, issues, and compliance

The approach is quite straightforward. Let us go through each step, using a case study to guide us. The case study involves a single department; in this case, an Art Department. The Art Department creates artwork for orders received from customers by way of the Customer Service Department, which precedes it in the overall process flow. As you would expect, the variability associated with this creative process added a level of complexity to the situation that needed to be factored into the design of the pull system. Further, not *all* orders required artwork to be developed; some customers provided their own. These orders would need to bypass the Art Department and proceed down a different path.

Step 1: Identify the Locations Where Queues Are Expected to Form

Again, orders are currently received by the Art Department from Customer Service. An art director quickly assigns the order to one of six graphic artists. In practice, that means that we have seven queues: one for each graphic artist, of assigned work; and one for the art director, of unassigned work. To simplify the process, the organization decided to maintain a single queue of unassigned work and have the graphic artists "pull" the work when they were ready. The art director would only have to monitor a single queue in order to determine how well (or not) the department was keeping up with demand. This approach also provided improved flexibility, since previously the art director had to reassign orders from time to time based on how individual graphic artists were doing on a particular day. With graphic artists "pulling" work only when they were ready to start it, the need for reassigning orders was eliminated.

Step 2: Identify Means to Provide Visibility

Once a single queue was established, this step was straightforward. Orders were received in "job folders" that contained all required information, on a table set up to receive the folders. It became quickly apparent that the department needed a way to categorize the orders by complexity. For the Art Department, complexity was driven mainly by the number of "elements" that had to be designed. Some orders only had two elements; others had eight. A simple review of the order easily determined the number of elements needed. Therefore, visibility of the number of elements, rather than the number of orders, was the key. The group agreed to use a categorization method based on the number of elements (i.e., 2, 4, 8), supported by color coding.

The department also needed to make visible the amount of time that the order was in the queue. The department's goal was to process received orders by the end of the next business day. For example, if an order was received on Tuesday, it had to be processed by end of business on Wednesday, or it would be considered late.

The department designed a board and located it adjacent to the table where the folders were received (see Figure 4.10).

There are three columns. One is by "category," and relates to the number of elements that an order requires (i.e., CD2, CD4, CD8). Two columns correspond to the day that the order was received. Because the department's

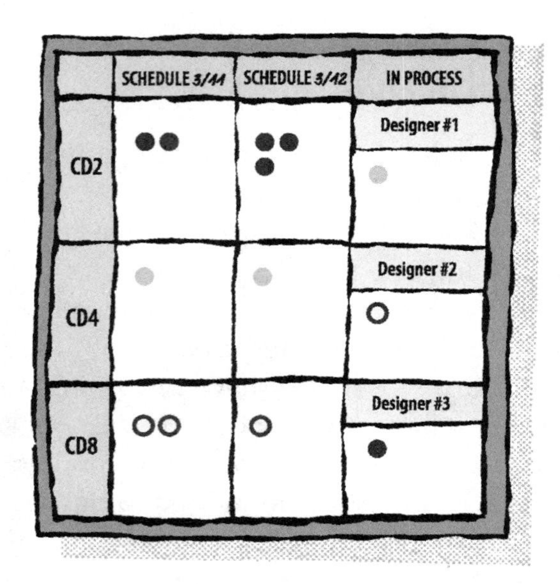

Figure 4.10 Art department board, partial.

goal is to complete all orders by the end of the next business day, only two days appear on the board. One column is for orders received yesterday, and the other is for orders received today. At a glance, anyone could see what the queue of orders and elements was at any time, and how long the orders have been in the queue. The department used color-coded magnets to enhance the visibility of the system, with a different color for each category.

On the right-hand side was an area for in-process orders. When a graphic artist pulled the next order, he or she would move a colored magnet corresponding to the order's category from the appropriate column displaying the queue (always taking from yesterday's column before today's) and placed in the area displaying in-process orders. This maintained both the visibility of the in-process orders and the orders in queue, thereby providing a complete picture of the workload at all times.

Step 3: Establish Limits for the Queue

The limit in this case was established based on the goal of the department. Remember, the department wanted to process all received orders by the end of the next business day. The board provided daily, clear visibility on whether or not this goal was being met. At the end of the current day, if any order remained in queue in *yesterday's* column, then the department was not keeping up with demand. This would trigger the "pull" of additional resources to match demand.

Step 4: Define Rules for the Queue

The board was reviewed at the end of each business day. If any orders remained in yesterday's column, the art director reallocated resources. The art director himself or herself could help out, or staff from another department could be reassigned the following day, since the company had people with graphic arts abilities working in other departments. With a clear understanding of the number of elements in the queue, an assessment could easily be made regarding the number of people that should be "pulled" the next day. Fortunately, it was very often the case that when demand on the Art Department increased, the demand on other departments decreased, because not all orders required artwork to be developed.

Step 5: Train People in the Pull System

The department created procedures for the pull system and trained its staff in them. The procedures were posted adjacent to the board shown in Figure 4.10. The training involved the use of the board, the color-coded magnets, etc. The ways in which the graphic artists performed their activities was unchanged; the change was simply in the way the process was managed. Consequently, the art director was most impacted by the new system, and he or she was instrumental in designing the system throughout.

Step 6: Monitor the System for Effectiveness

The art director periodically audited the system to make certain that the board and color-coded magnets were being used properly. The procedures were reinforced during morning "huddles" at the board with members of the department. The department decided to add some performance metrics to the board to provide a more complete picture as to how the system was working. The number of "late" orders and overall throughput were both tracked (see Figure 4.11 for the "complete" board, which will continue to evolve over time). On the right-hand side, the procedures and metrics were posted.

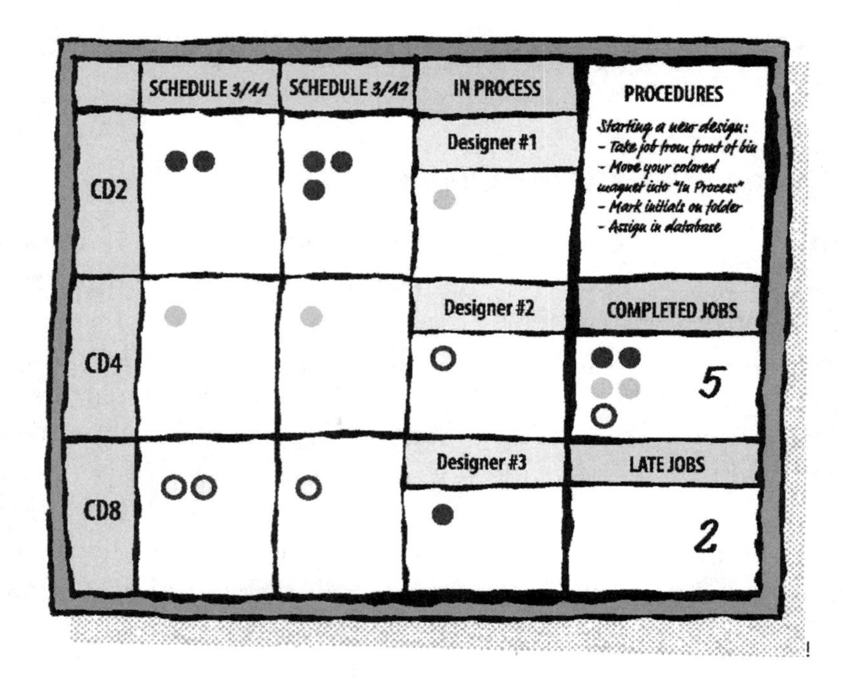

Figure 4.11 Art department board, complete.

Benefits of Office and Service Pull Systems

What are the benefits of implementing pull systems in office and services? One benefit is greater predictability in the lead time for a process. Since the amount of work in process at anytime is controlled, the lead time through the process becomes consistent and predictable, which can help you consistently maintain customer satisfaction.

Another benefit of implementing pull systems is that it makes the management of the process in question easier. Since the system is most often worker managed, process or operations managers find that their traditional roles will change. They move from a "traffic cop" role—directing every point of the process—to managing a single point of the process. Typically, this point is a queue at or near the beginning of the process. And most often, the key responsibility is to monitor incoming demand and compare it to the capacity for which the overall process is designed (see Figure 4.12).

Management can also focus more on effecting process and business improvement. This important activity often goes unattended, as managers get caught up in the day-to-day direction of even the most basic tasks of

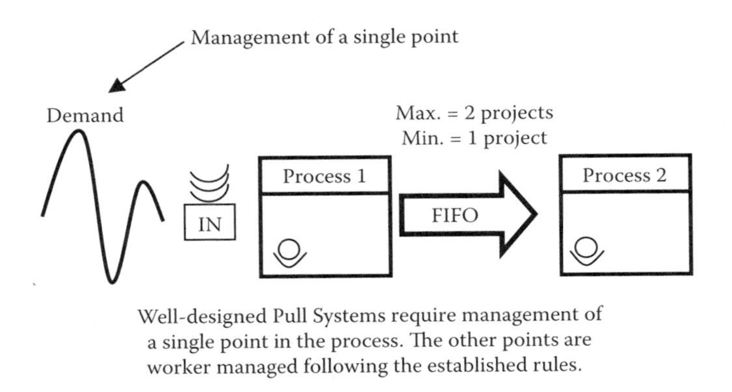

Figure 4.12 Single point of management.

most work environments rather than focusing their attention on the big picture. We will discuss this point in more detail in Chapter 8.

Further, when you implement Lean in office and service environments, the increased flexibility you will see will result in improved use of resources. Through cross-training, you will find that you are using people's full talents and abilities rather than just a small portion of them, thereby maximizing the benefits of the pull system. Unfortunately, there is a serious *lack* of cross-training in most office and service environments, most often ascribed to a lack of time to cross-train. Interestingly, implementing pull systems can drive cross-training, as staff will no longer be overproducing; consequently, time is freed up for this important activity.

A financial institution manager effectively used pull signals to identify opportunities to provide cross-training for the associates in her department. She implemented a pull system throughout the department for the various information processes in which her associates were involved. Then, when an associate was "triggered" to move to another activity, she took advantage of this opportunity to deliver training to that individual on an as-needed basis. She also used this opportunity to conduct a kaizen event, or a focused improvement effort to streamline a particular processes or steps within a process. Once the streamlining was completed, the new process was documented in a simple way, through standard work instructions, and the associate was then trained on the new process.

Over a 3-month period of time, she conducted dozens of kaizen events, resulting in streamlined processes with process times reduced up to 30%, documented processes, and a fully cross-trained work force. All of these improvement efforts were driven by the pull system itself. The associates in her department welcomed the opportunity to learn and perform different tasks, and found the work environment to be more satisfactory and rewarding as a result.

Summary

Remember, you will often want to implement pull systems prior to implementing flow systems, because pull systems can quickly provide much-needed stability and predictability to the existing system. Further, the cross-training inherent in pull systems provides an excellent foundation for the implementation of the flow concepts described in Chapter 3.

Most office and service organizations are just scratching the surface when it comes to applying pull systems in their operations. However, with a deeper understanding of the concepts, the opportunities to apply them will become more apparent. With a solid understanding of the basic elements of all pull systems, people will be better able to adapt the concepts to successfully implement them, and realize the benefits aforementioned. All of the examples provided in this chapter represent creative ways that organizations have adapted pull concepts to meet their unique needs. In the next chapter, we will expand on the concept of "Visual Management," a critical part of the pull systems described in this chapter. There are many other applications for Visual Management that can provide important benefits to an organization.

Chapter 5

Establishing Visual Management in Office and Services

Implementing visual management techniques is a critical component of any Lean enterprise. Why is visual management so important, and what does visual management really mean? In this chapter, I will answer both questions and provide you with plenty of examples along the way.

Background

It is important to understand the reasoning behind the use of visual management techniques. Humans tend to be very visual creatures, making visual communication very efficient and effective. Seeing really is believing, and often, a picture really is worth a thousand words. Further, retention of information decreases with time—*any* amount of time. Ever hear the expression "out of sight, out of mind?" More than 60 years ago, György Kepes, founder of the Center for Advanced Visual Studies at the Massachusetts Institute of Technology, in his book *The Language of Vision* (Chicago: Theobald, 1944), noted:

> The visual language is capable of disseminating knowledge more effectively than almost any other vehicle of communication. With it, man can express and relay his experiences in object form. Visual Communication is universal and international; it knows no limits of tongue, vocabulary,

or grammar, and can be perceived by the illiterate as well as the literate. Visual language can convey facts and ideas in a wider and deeper range than almost any other means of communication.

And visibility provides other benefits, including an improved probability of sustaining standard work practices; a greater sense of belonging, accountability, and pride; and increased customer and supplier confidence in the organization.

This is not to say that there are no challenges to applying visual management techniques in any work environment, and in office and services in particular. We have touched on these aspects in earlier chapters, but first and foremost is a general reluctance to accept transparency in the workplace. Why this reluctance? There are several possibilities. Often, reluctance to make work visible can be traced to a fear of an unsupportive or even punitive response. This is particularly true when attempting to make *performance* visible. People are often reluctant to make such information visible if there is an organizational history of punitive responses to failure to meet performance expectations.

However, it is not only a fear of potential consequences for not meeting performance goals that can drive resistance to making work visible. Another source of reluctance stems from the ever-present belief that information is power. Fears about job security and position in the pecking order can also lead to a reluctance to make standard work visible, since other employees can learn how their co-workers perform particular activities. "Tribal knowledge"—knowledge that one or two persons have, but which is not adequately shared with others—will often be documented as "key points" in standard work, which can be threatening to those who believe that job security lies in being the only person with specific knowledge or skills. You can expect strong resistance to posting standard work from those caught up in job security fears.

Therefore, it is critical that top management allay these fears prior to implementing visual management techniques. You need to convey a strong message that visibility and transparency *will* be a part of the Lean system, and that resistance will not be tolerated. Although you can be flexible in *how* you make work visible, the underlying concept cannot be compromised.

Of course, this means that management itself must buy into the concept, and therein is the challenge. Most resistance to visual management concepts comes from management, particularly middle management. Often, they will argue that the information employees need is contained elsewhere (in the computer, for example). Managers who could not tell you the color of the walls in the office will suddenly become extremely averse to "cluttering them up." Although these are not valid arguments, they can still bring the effort to a screeching halt.

Approaches to Visual Management

You can make work visible by a number of means. In general, the approach you will choose depends on two key factors: proximity and complexity (see Figure 5.1).

Proximity is the relative location of the provider of information—the person or persons who possess the information—to the recipient of the information—those who have some need for the information. You can use simple methods to make work visible, such as whiteboards, as long as the provider and recipient are physically proximate or located close together. When the provider and recipient of the information are separated by a great distance, then you may want to consider using electronic techniques (e.g., computer-based software applications). Since necessity really can be the mother of invention, some companies have used creative methods to overcome the proximity issue; for example, taking a digital photograph of a whiteboard and e-mailing it to another location so that boards in both locations are kept up to date and remain "matched."

Complexity refers to the amount of information to be communicated. For example, whiteboards can be very effective for displaying project statuses if the number of projects in process is limited, say, to less than 50. When projects in process exceed that number, however, organizations often turn to electronic methods.

Although electronic techniques can be effective communication methods, they do not offer some of the other benefits inherent to making work visible (e.g., improved probability of sustaining standard work, greater accountability). I am sure that you can guess why electronic techniques are not

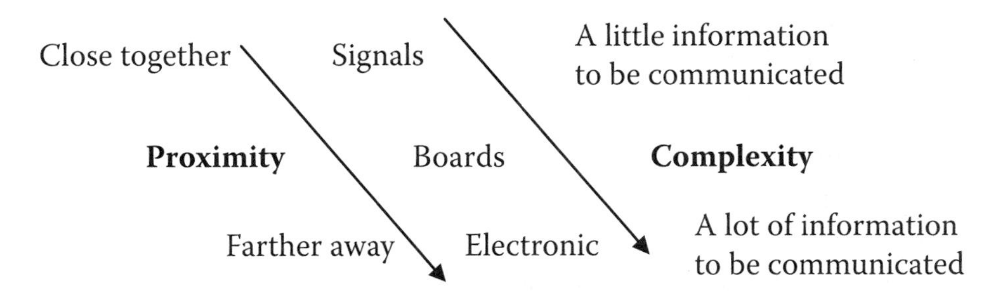

Figure 5.1 Factors impacting visual management approaches.

as visible in the workplace as are other methods. One way to address this shortcoming is to install monitors in appropriate locations in the workplace to provide improved visibility.

At one company, phone system software provided real-time statistics on wait time, dropped calls, etc., by "call loop." However, the full benefits of this information were not realized until it was made available to all of the associates rather than just on the supervisor's desktop computer. By using monitors, associates knew when to lend assistance to each other with little or no direction from their supervisor. Performance significantly improved as a result.

When determining an appropriate approach for making work visible, always remember the "keep it simple" rule. Too often, organizations put great effort into developing more complex techniques when simpler methods would work as well, and could be implemented in less time and at a lower cost. Lean thinkers always look for effective approaches that can be implemented quickly. You also need to consider ease of maintenance of the implemented methods. Any technique that requires substantial effort to keep up to date will not be maintained over time. For example, if your employees need to constantly rewrite the same information on a whiteboard, they will quickly grow frustrated and will stop using it. However, what if you use dry-erase magnets on the board? Then the information could be quickly reorganized with minimal effort.

The same is true of electronic techniques. If associates are required to enter information into a software application strictly for the purpose of maintaining the visual management system, they are likely to grow frustrated over time. If the necessary information is already in the system and the challenge is in extracting it, techniques can probably be developed to do this easily, perhaps using a report generator.

Elements of Visual Management

To determine the elements of visual management needed for your specific workplace, you will need to ask yourself a series of questions. First, let us take a look at the questions:

- What is the purpose or function of the area?
- What activities are performed in the area?
- How do people know what to do?
- How do they know how to do it?
- How do they know how they are doing?
- What is done if performance expectations are not being met?

Now, we will explore each question in depth and determine how to make the *answers* to each question visible within the workplace.

What Is the Purpose or Function of the Area?

There are a number of benefits to making the purpose and function of an area visible. The most obvious benefit is to new employees or employees who are new to an area, as making the purpose and function of the area visible will help them to become oriented more quickly to their new work environment. As a result, they are likely to have fewer questions about the work, which can often disturb other employees. However, the benefits are not limited to new employees; longer-tenured employees in larger organizations also often struggle with a lack of proper awareness of the locations and functions of different departments. Although the benefits might seem minimal, making an area's purpose and function visible is a quick, easy fix, and worth it for the potential to minimize disruptions. Typically, a simple sign posted prominently will suffice.

What Activities Are Performed in the Area?

In addition to making the purpose and function of an area visible, you will also need to clarify which tasks are performed in any given area. You can do this by posting value stream maps in the area to define the overall process and to highlight the area's particular role in it (see Figure 5.2).

Another technique to visibly convey this information is a Supplier-Input-Process-Output-Customer, or SIPOC. This single-page form succinctly displays this important information (see Figure 5.3).

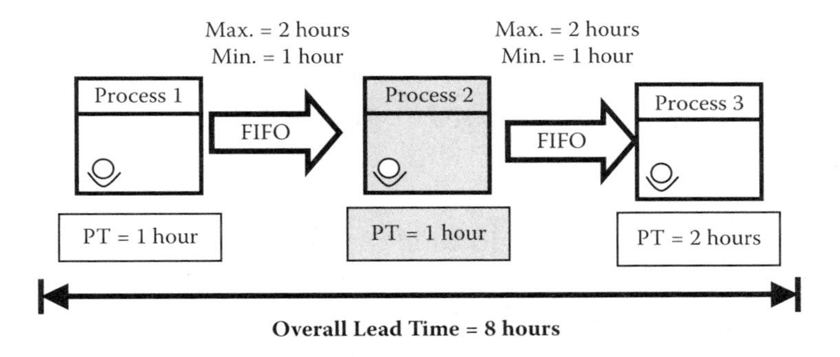

Figure 5.2 Using a value stream map to describe the overall process with a particular area highlighted.

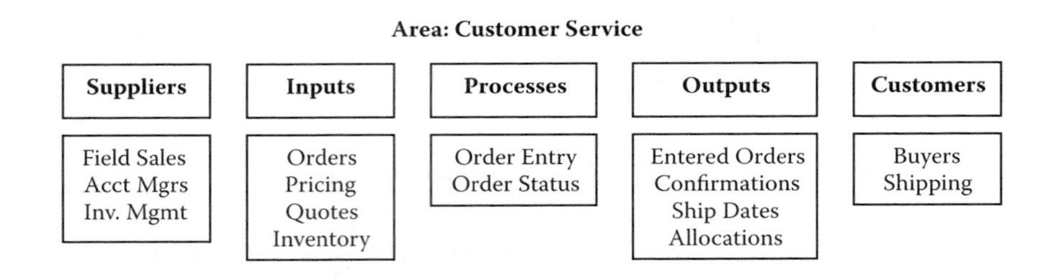

Figure 5.3 Example of a SIPOC.

How Do People Know What To Do?

In other words, how is work scheduled, triggered, and prioritized? This question usually leads to very interesting discussions, since there is often a broad range of ways in which work is managed within organizations. Often, the organization's management will provide such direction, even for basic tasks. In other instances, the people performing the work make these decisions. As we discussed in earlier chapters, however, they do so in a way that is *not* best for the overall process, and some form of scheduling system needs to be put into place. In project-oriented environments, schedules that clearly display priorities can simply be posted in the workplace, perhaps using whiteboards. Another option, gaining in popularity, is to create an "obeya" or "big room." An obeya is simply a room that displays various visual project management techniques to help manage complex projects such as product development projects. These simple types of visual project management techniques have been used for decades. Figure 5.4 provides an example of a project scheduling board.

The concept can be expanded beyond simple project scheduling. The example in Figure 5.5 conveys "voice of the customer (VOC)" information and has areas for different functions such as Design and Sourcing to communicate with other functions. This represents a comprehensive communication system to manage projects such as those in Product Development.

Another option for visually displaying the work to be completed is a "plan for every process" (refer back to Figure 1.3 in Chapter 1 for an example), which can be very effective in multitasking work environments in which activities are regularly performed with a fixed frequency (e.g., daily, weekly). In a plan for every process, a schedule for performing activities is agreed upon and posted, often as part of standard work developed for particular roles in the office or service. The schedule is coordinated among roles to ensure that overall process performance expectations are met, as

Project	Order Entry	Generate Drawings	Bill of Material	Materials On Order	Receive Materials	Production Start	Production Complete
ABC	COMP	9/1	9/3	9/4	9/21	9/21	9/28
DEF	COMP	9/2	9/4	9/5	9/28	9/29	10/5
GHI	COMP	COMP	COMP	COMP	9/20	9/25	10/2

Figure 5.4 Project scheduling board.

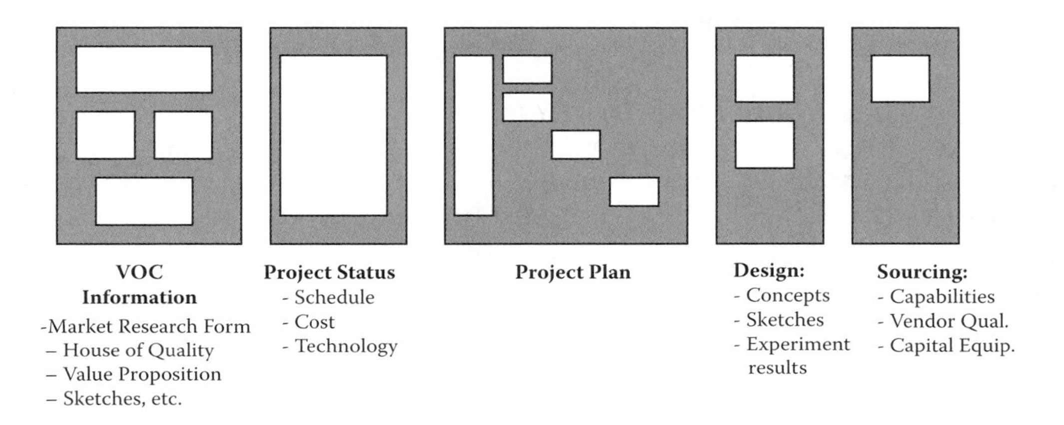

Figure 5.5 Comprehensive visual project management system.

Time	Monday	Tuesday
9:00–9:30 a.m.	Check e-mail ●	Check e-mail
9:30–10:00 a.m.	Enter orders ●	Enter orders
10:00–10:30 a.m.		
10:30–11:00 a.m.	Unscheduled work ●	Unscheduled work
11:00–11:30 a.m.	Process RMAs ○	Work on order holds
Etc.	Etc.	Etc.

Figure 5.6 "Plan for every process" with visual indicators.

I described in Chapter 1, and is posted in a designated area, and/or at each person's workplace (e.g., office, cubicle). Evidence that tasks were completed at designated times is usually included as part of the visual management system. The example from Figure 1.3 is shown in Figure 5.6 with visual indicators added to show whether each task was completed at the appropriate time. A green dot could be added to indicate that a task was completed on time. A red dot could indicate that it was not.

In Chapter 4, we discussed how to implement pull systems in office and service environments. Pull systems can be used to trigger the completion of particular activities. After all, a pull system is a form of scheduling mechanism. All rules associated with the pull system must be posted in the area for all to see. In Chapter 4, I provided a number of examples that showed how simple signs can be used to clearly display how people should prioritize and schedule their activities.

How Do They Know How To Do It?

In Chapter 2, we discussed the concept of standard work. As you will recall, standard work defines the steps that must be completed in order to perform a process or operation. Standard work also includes "key points" that define "how" to perform each step. Key points include *clarity affecting efficiency* (i.e., perform the step in this way to ensure speed of completion), *quality* (i.e., perform the step in this way to ensure quality of completion), and sometimes *safety* (i.e., perform the step in such a way as to ensure its safe completion). The expected time needed to complete an operation as well as the specific timing of completion where appropriate are also included in standard work.

Standard work is typically displayed on a single-page document posted in the area where it is performed. As you will recall from Chapter 2, people will drift away from standard work procedures over time unless there are visual reminders to sustain them. To realize the benefits of standard work, it is imperative to keep it visually front and center. Posted standard work is a reminder to people of how to perform the work to ensure that the required performance levels are met (e.g., throughput, quality). If standard work remains in electronic form or hidden in a book or binder, ensuring that it is followed over time becomes progressively more difficult.

Remember, another purpose of standard work is to allow for nonstandard conditions to be more readily identifiable. Process time is included in standard work. Let us say that the associate performing a process or operation glances at the process time visually displayed on the standard work document and sees that it should take 10 minutes to complete the process but realizes that it is taking 20 minutes in this particular instance. This is a nonstandard condition and, ideally, should be identified by the person performing the task. Why is it taking longer to complete the task than it should? A little investigation may reveal that some undesired change to the process has occurred and should be acted upon. Of course, the converse is also possible.

Let us say that the process is taking 5 minutes when it should take 10 minutes. Perhaps the associate is taking some shortcuts—in other words, is not following the standard work—which may result in problems during a subsequent process. Again, unless it is posted visibly, this information will be out of sight, out of mind.

How Do They Know How They Are Doing?

In general, people want to know how they measure up. That is why scoreboards are maintained in most sporting events. Why not provide the same feedback and performance visibility in an office and service environment? The answer is that we can, if people are willing.

In Chapter 3, we reviewed an example of a "Plan versus Actual" or "Pitch" board (Figure 3.7), which displays actual output versus expectations. Pitch boards are typically worker managed, clearly displaying the goals at some frequency (the pitch), and how well an individual or team is doing in meeting the goals. The comments for any period of time when the goal was not met are *very* important. Periodically reviewing the board and comments will help to identify recurring problems to be addressed to ensure performance in the future. These techniques have been effectively used in manufacturing, and are just now finding their way to office and service environments.

Performance over longer periods of time can be tracked by posting process metrics periodically (i.e., daily, weekly, monthly, etc.). Key performance measures or indicators, often referred to as KPIs, can be posted on a simple dry-erase board or some other media. "Traffic light" techniques, using colors to quickly display which measures are meeting expectations and which are not, can enhance the effectiveness of whatever technique is used. A green mark next to a metric indicates that the individual or team is meeting performance expectations, while a red mark indicates the opposite (See Figure 5.7). Of course, the key here is to select the correct measures that will affect the desired behavior and response. We will talk about this in more detail in Chapter 8.

It is not just measurements that indicate current performance. Another indicator is whether or not various required tasks are being performed when they were needed or planned (see Figure 5.8). Figure 5.8 tracks completion of individual tasks over the course of a week, but could easily be adapted for an entire department if desired. Similarly, it could be adapted for a day or a month, whatever is appropriate. This visual indicator can be as simple as a small board located at or near the person's work area.

Metric	Goal	Actual	Status	Comments
Sales Revenue	$4M	$3.8		
On Time Delivery	98%	98%		
5S Score	90	85		
Etc				

Figure 5.7 Performance Measurement Board.

Monday	Tuesday	Etc
General Ledger entries ●	Complete aging report ●	Etc
Issue credits ●	Prepare for management meeting ●	Etc
Complete special project ●	Set-up new customers ●	Etc
Etc	Etc	Etc

Figure 5.8 Visual indicators of the completion of tasks.

In it, we see nonrepetitive activities such as "prepare for management meeting" and "complete special project." In the case of the management meeting, there is a clear due date, as the meeting has been scheduled for a particular day. However, the due date for the special project was not clear. The service provider established Monday as the day to complete the project. Color coding can be used to distinguish repetitive tasks (in black) from nonrepetitive tasks (in some other color). Color coding also displays the status of each activity. Red indicates that the activity was *not* completed on time, whereas green indicates that it was. These simple, easily maintained techniques allow the organization to determine whether or not the demand on resources is presently being met. This ability to monitor capacity versus demand on an ongoing basis is very important in most office and service flow systems.

Similar techniques can be used in project management environments to display whether key milestones have been met or are in danger of not being met or were missed. A green mark can indicate that the project is on schedule, a yellow mark can means that the due date is in danger of not being met, and red can indicates that the date will be or has been missed. I have adapted Figure 5.4 to demonstrate how this might work (see Figure 5.9).

The four techniques suggested here (standard work, pitch boards, performance measurements, and task completion boards) will provide employees

Project	Order Entry	Generate Drawings	Bill of Material	Materials On Order	Receive Materials	Production Start	Production Complete
ABC	COMP	9/1 ●	9/3 ●	9/4 ●	9/21	9/21	9/28
DEF	COMP	9/2 ○	9/4 ○	9/5 ○	9/28	9/29	10/5
GHI	COMP	COMP	COMP	COMP	9/20 ●	9/25	10/2

Figure 5.9 Project Status Board using traffic light techniques.

with a comprehensive picture of their performance, both in the short and long term.

What Is Done If Performance Expectations Are Not Being Met?

Knowing how we are doing is one thing, but we also need to know how to respond and make adjustments to the work when necessary. The Lean concept of "jidoka," originating in the early 1900s, refers to providing people with the ability to detect abnormal conditions and to immediately stop work. An abnormal condition can be a failure to meet performance expectations or a quality problem. In most office and service environments abnormal or nonstandard conditions continue unaddressed with employees sometimes even exhibiting pride in "getting it done" in spite of the problems. Although we can admire their tenacity, this approach is completely unacceptable. Again, standard work can help us identify abnormal conditions and issues with performance expectations.

Now, not all problems warrant stopping work. However, at the very least, the problem needs to be addressed while the work proceeds. Therefore, guidelines for soft or hard "line stops" need to be clearly defined and posted in the area. In a "soft stop," employees make the appropriate people aware of a problem, and then continue with the process. A "hard stop" means that the process will not be permitted to continue. Soft and hard stop guidelines need to include a desired response time, or at least a date when the problem is expected to be addressed.

Responses can be short term (e.g., call someone for help), intermediate (e.g., re-train an employee who is not following standard work), or long term (e.g., some process improvement project). Further, responses can, at least in part, drive the continuous improvement efforts within a department (more on that later). Therefore, they are often posted as part of a continuous improvement component of the visual management system. Again, a simple dry-erase board, on which recurring problems and follow-up actions are identified, will suffice.

You cannot allow problems to go unattended. Therefore, an "escalation" process is often included with the list of problems or follow-up actions. The escalation process triggers the involvement of others, typically higher-level management, to help complete the follow-up action. Let us take a look at an example of a project board that includes an escalation process (see Figure 5.10). Of course, the board can take various formats, and the specific escalation process will depend on the organization. The board should be reviewed periodically to make sure that the department or team and others in the organization are following through on problem resolution.

Including Continuous Improvement in Visual Management

Now that we have run through the questions you need to answer to identify the key elements of visual management systems, you should have a sense of how to proceed in implementing your own system. But there is another critical element of visual management systems: continuous improvement. No matter how well you have implemented Lean within your organization, there is *always* room for improvement, and you need to keep those goals visible. Remember, out of sight really is often out of mind. Your continuous improvement information might be Kaizen or Rapid Improvement Event Planning Sheets, one-page documents that describe an upcoming rapid improvement event, its objectives, scheduled dates, team members, etc. A Kaizen, or Rapid Improvement, Event involves an identified team working for a designated period of time to implement improvement in a very specific area. These events are planned in advance, and preparations are made so that the team will be able to identify and actually implement changes within the designated timeframe. They are very focused improvement efforts that result in real change.

It could be an "A3" or "storyboard," another one-page document that conveys a story about a problem-solving or process improvement effort either in process or completed. "Storyboards" were first used in the 1960s but became more common as part of the Quality Management movement in the 1980s. Teams of associates working on an improvement effort would use them to communicate their objectives, the direction they were taking, the progress that they had made, and the results achieved. In addition to posting information about continuous improvement efforts already in process, you can post information that can *identify* and *drive* opportunities for improvement as well. For example, the findings from a Workplace Organization or "5S" audit, which we will discuss in detail in Chapter 6, can be posted pending review

Problem	*Action*	*Responsibility*	*Completion date*	*Status*
1. People not following standard work	Re-train	Supervisor	October 5	OK
2. Address computer problem	Identify problem to IT. IT to correct	Manager	September 3	Escalation Level 1
Etc.				

Escalation Level 1: Get appropriate director involved.

Escalation Level 2: Get leadership team involved.

Figure 5.10 Problem board with escalation process.

and follow-up by the area team. The same thing goes for findings during "gemba walks," which we will explore in Chapter 8. All findings could be added to the Problem Board (see Figure 5.10), serving as a central collection point for all improvement ideas.

Summary

Together, all of the elements described in this chapter constitute a comprehensive visual management system. This type of system can provide important benefits to your organization, including:

- Greater stability and predictability of a process, department, or an entire organization
- Less overall effort to manage the process
- Greater awareness of problems and opportunities for improvement
- Greater responsiveness to problems
- Improved communication and awareness for all employees
- Shorter learning curves for new or reassigned employees

Visual management can be a tough sell to people who have no previous experience in such practices. It almost requires a leap of faith. Trust me when I tell you that you would not be spending half of your day updating whiteboards if done properly, the responsibility for maintaining the various elements of the visual management system is spread among team members. Remember, Lean concepts are *team* concepts and are best applied through employee involvement.

Chapter 6

Lean Tools for Office and Services

In Chapters 1 through 5, we reviewed the foundational concepts of Lean: Value and Value Stream Management, Standard Work, Flow, Level Pull, and Visual Management. When put into practice, these concepts fundamentally change how work is performed, how it flows, and how it is managed. In addition, a number of Lean tools can be applied in office and service environments to further improve performance, or sustain the gains you have realized. These tools should be applied *in support* of the Lean key concepts we have already covered, not in place of them.

It is important to remember that tools are a means to an end, not the end themselves. Typically, the isolated application of most tools does *not* fundamentally change how the work is performed. Too often, organizations miss or gloss over this important point. Often, the Lean office or Lean service effort consists of the uncoordinated application of various Lean tools. I can guarantee you that these companies will not realize the significant Lean benefits that we have covered in earlier chapters.

Nevertheless, these tools can help you reach a higher level of performance and, as such, have a place in any Lean office or service environment. In this chapter, we will review the most common and effective Lean tools for office and service environments: Workplace Organization (also known as "5S"), Mistake Proofing, and Setup Reduction, or Quick Changeover. Before we get started, note that there can be some overlap between Lean foundational concepts and these tools. For example, some would argue that Workplace Organization is a foundational concept of

Lean, and that may be true—but only *if* it is properly applied. For our purposes, we will consider it a tool.

Workplace Organization or 5S

Workplace Organization (often called 5S—more on that shortly) is arguably the most widely used Lean tool in office and service environments. Unfortunately, with its widespread use comes widespread *misuse*, with many companies viewing it (incorrectly) as nothing more than a safety and housekeeping tool.

The objective of workplace organization is to create a *functionally* organized work environment, and not to simply give the *appearance* that things are organized. Many companies implement 5S concepts, and the workplace gives the appearance that it is highly organized. Unfortunately, people who work in the area are not working in a more effective or efficient manner. In other words, the workplace has not been functionally organized.

What is the distinction between superficial and functional organization? Let us take a look at how most organizations audit their workplace organization systems. The audits often do not involve much more than inspections of the area: that is, have items been labeled by location, and are people returning these labeled items to their designated locations? Has the area been cleaned at the end of the workday? These audits miss the mark in that they do not observe processes as they are being performed. Are people *working* in a functionally organized way? In other words, it is less important to confirm that people are returning items to their designated locations than it is to confirm that they are the best locations possible for the items.

As I mentioned a little earlier, workplace organization is also known as 5S (see sidebar). The S's we will use here are

- Sort: Identify unnecessary items that are found in an area.
- Set-in-Order: Place items in the best locations possible.
- Shine: Maintain an area in good working condition (clean and safe).
- Standardize: Policies, procedures, practices to maintain the first three S's.
- Sustain: Discipline to the first four S's.

There are numerous versions of the 5S's, such as Straighten instead of Set-in-Order, but those differences are insignificant. Together, the 5S's provide a methodology for getting and staying organized.

SIDEBAR Origin of the 5S's

The origin of the 5S's can be traced to five Japanese terms, each beginning with an S sound, describing workplace practices conducive to visual control and Lean production:

Seiri: Separate needed from unneeded items—tools, parts, materials, paperwork—and discard the unneeded.
Seiton: Neatly arrange what is left—a place for everything and everything in its place.
Seiso: Clean and wash.
Seiketsu: Cleanliness resulting from regular performance of the first three S's.
Shitsuke: Discipline, to perform the first four S's.

The terms have been translated to English. The meaning of these five terms has been expanded so that people understand their purpose and the depth to which they must be applied. The expanded definitions bring them more in line with the true meaning of their Japanese counterparts. For example, in Shine, we include the concept of "inspect through cleaning," which goes beyond the simple translation of Seiso (clean and wash).

At first glance, the 5S's do not seem too complicated, but there is a lot of depth to each "S." Let us start with "Sort." Again, remember that our goal is functional organization, so it is less important to identify unnecessary items within an area than it is to identify the root cause of the item being out of place to begin with. Maybe in this case, a designated location was never identified for the item. That is a failure in the second "S," Set-in-Order. Or maybe the designated location for the item is wrong; in other words, it is not easy to retrieve or return the item to that particular location. The solution is simple: the item should be relocated.

Your sorting process should be ongoing: everyone in the organization should be willing to identify items that seem out of place in a timely manner. Sorting should *not* be an isolated event, taking place several times a year, as is often the case. The ongoing process of sorting is called *red tagging*, and a defined red-tagging process is part of every 5S system. Red tagging provides a "standard" (the fourth "S") process for sorting. Applying a red tag to an item poses the question, "why is this item here?" A red tag does not mean that the item will be disposed of. It is critical to ensure that staff understand this; otherwise, the likelihood of their following the red-tag process and policies will decrease. People will be reluctant to attach red tags, fearing that the item may be wrongly discarded. Again, make it clear that the red tag simply poses a question that others will answer.

Typically, a defined cross-functional team will review all items that have been red-tagged, performing the necessary investigation and making the appropriate decision. Your red-tag process needs to include a defined timeframe within which these decisions will be made. This prevents procrastination from impeding the process and items from accumulating within the wrong area. Make sure that everyone in the organization is familiar with the 5S system, including the red-tag process. In an office or service environment, sorting will include the physical items in an area, both electronic and hardcopy files, office equipment, office furniture, etc. Often, sorting cannot be initiated until sort standards have been established (more on that later).

The second "S" is Set-in-Order. Our key objective here is to identify the best location for every item. This "S" improves functionality. Too often, people identify incorrect locations for items, often locating them too far from the point at which they are used, creating excessive motion waste. At other times, items are located in inconvenient locations that require turning, twisting, lifting, walking, etc. If an item's assigned location is inconvenient, the likelihood of it's being returned decreases. During the Set-in-Order activity, you will test locations to verify that they are the best available.

Relocating or even purchasing items is also part of the Set-in-Order activity. For example, at one company, a central multifunction copier/fax/printer was used by office personnel to scan documents for e-mailing or archiving purposes. This required each person to leave their work area every time they needed to scan documents, as frequently as ten times each day. The organization's solution was to purchase and install low-cost scanners at each associate's work area, linking them to his or her desktop computer. Consequently, the need to get up and walk to and from the central copier was eliminated.

Another component of the Set-in-Order activity is labeling items and locations for items, also called *home addresses*, including electronic files and folders on servers and desktops. However, you cannot initiate this activity until Set-in-Order standards have been established (more on that shortly). For example, before you can label files and folders, you need to develop standards for file and folder names. Often, the greater benefit is not derived from simply identifying locations and labeling items, but rather from the standards that have been established across the office or even across the organization.

The third "S" is "Shine." One objective of this "S" is to identify the most effective and efficient methods of cleaning. Obviously, there is more benefit to this objective in a production environment. Another objective of Shine is to implement countermeasures or methods to prevent problems from occurring. In an office environment, this can involve straightening computer cables, phone lines, etc., to prevent equipment damage, eliminate safety hazards, and to make it easier to clean particular areas. Ensuring sufficient lighting and safety are additional components of "Shine."

In general, "Shine" does not provide significant benefits to people working in office or service environments (although there are exceptions). However, it should be a requirement of your workplace organization plan nonetheless. People need to feel comfortable and safe in their work environments, and studies have clearly shown that people's performance is directly correlated to the condition of their work environment. For example, studies conducted from 1924 to 1932 at the Hawthorne Works facility outside Chicago, Illinois, determined that lighting, maintaining clear workstations, clearing floors of obstacles, and related attributes resulted in increased productivity for periods of time. Of course, in many service environments such as health care and food services, "Shine" is an absolute necessity. In other services, the condition of the workplace can determine whether a person becomes your customer, or goes to another service provider.

The fourth "S" is for "Standardize"—in other words, establishing standards for the first three "S's." This is where the real benefits of 5S in the office or service are realized. This S takes us back to the need for standard work, since it can be nearly impossible to organize any area, shared or otherwise, if everyone works in substantially different ways.

For example, a common "Sort" standard in offices focuses on documentation retention, either electronically or in hardcopy. Different people often follow different rules or standards for file retention; some keep everything forever, and others discard things too quickly. However, developing a Sort standard is made easier because there are local, state, and federal requirements for file retention standards for key business documents. Be sure that your organization is familiar with them and establish Sort standards to meet these requirements. Unfortunately, there are a myriad of documents lacking these standards, and it is left to your organization to define your own. The first step is to determine practical needs for retention. At one company, quotes were kept on file for up to 2 years, though the quote was only valid for 3 months. It is also important to establish clear electronic file retention standards. At another company, the declining response time of the computer system was attributed to the amount of data maintained on the system—from as far back as 10 years. This organization developed archiving standards to alleviate the problem, keeping only 2 years of data. Simple methods to access older data were developed and provided to users in the case it was needed, which was infrequent.

You need to use methods that make sorting easy. Let us take a look at a situation that takes place all too often in office and service environments. One or several people have been assigned to carefully go through file cabinets to identify files that can be removed, perhaps because of the age of the files. Unfortunately, this important activity requires far too much effort, as each file must be carefully examined to determine its age. However, there are ways to identify the age of files without having to review each one in turn (see Figure 6.1). Using visual techniques, a task that previously took hours to finish can be reduced to mere minutes.

A key Set-in-Order standard is the 30-second test. You should be able to find any item within the workspace in 30 seconds or less. This rule applies to *everything*, from staplers to file folders to e-files. To meet this objective, you need to create standard file and folder naming conventions. You cannot adequately organize electronic and hardcopy files without such standards. You can even conduct a 5S kaizen event—a short-term improvement effort focused on improving workplace organization—specifically on the subject of electronic and/or hardcopy filing systems. This can significantly reduce the time you lose searching for information. In some cases, this can result in regaining as much as 10%–15% of a person's available time, equivalent to approximately 1 hour out of an 8-hour day. Who would not want to have an extra hour each day?

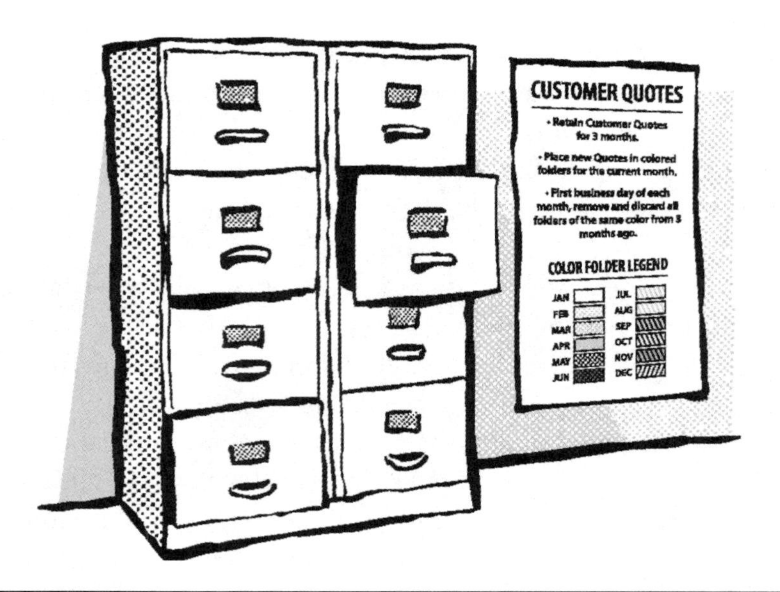

Figure 6.1 Example of a color file system.

Another Set-in-Order standard involves establishing quantity limits for supplies, using a simple supermarket pull system (see Figure 6.2). You can establish a maximum quantity limit for each item and provide sufficient space to accommodate the established quantity. More will not be purchased than the maximum quantity limit. By establishing a minimum quantity limit to trigger reordering of an item, you can ensure its availability. In fact, such systems are often implemented as part of an office 5S kaizen event.

Next, let us take a look at the "Shine" standard, which includes regular cleaning routines. In an office environment, regular does not necessarily mean daily. Perhaps weekly will suffice. In most service environments, however, maintaining appropriate conditions and ensuring a positive customer experience require a higher frequency. Whatever the standard, again, it must be made visible. Examples of visible cleaning standards can be found in hospitality industries, such as retail stores (e.g., signs reminding employees to wash their hands after using the restroom; cleaning checklists posted on the backs of restroom doors).

A standard at many hotels is to flip mattresses periodically to maintain comfort and to extend the life of the mattress. One hotel attached labels to

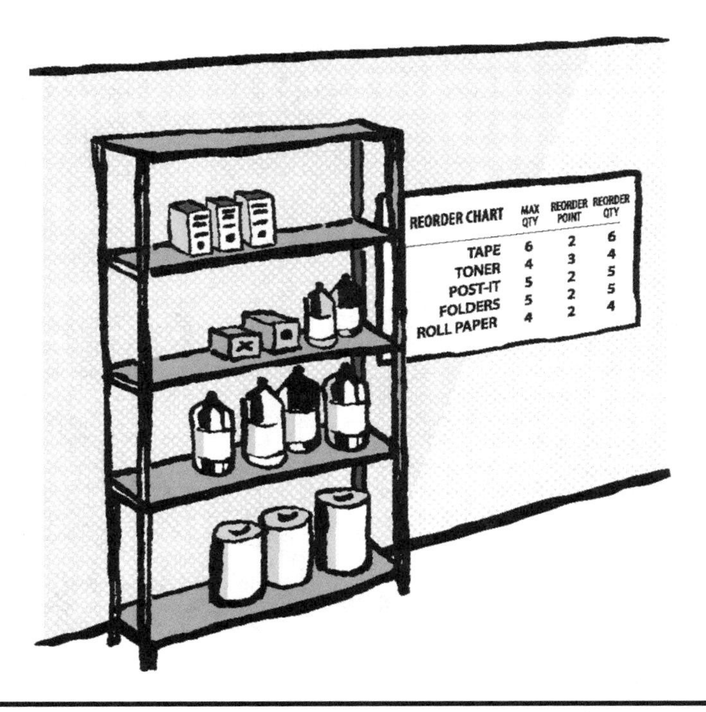

Figure 6.2 Supermarket pull system for office supplies.

mattresses to make the standard visible. The labels displayed the months that a particular side of the mattress should be facing up (or, conversely, down). Housekeeping personnel were reminded of the standard every day the bed was made (see Figure 6.3).

Once you have implemented the first four S's, you need to Sustain the fifth S. Every organization needs a Sustain model, which should include the following elements:

- Periodic audits of each area, using a standard checklist and scoring system
- Recognition for maintaining and/or improving workplace organization in particular areas
- Training of new associates in 5S (and Lean in general) so that they understand the system and their role in it
- The periodic application of the 5S methodology as part of focused 5S kaizen events

A standard checklist for use in workplace organization of 5S audits should be available that everyone understands and uses (see the appendix for an example). In addition, you need to define an effective audit *process*. Your audit process cannot be performed solely by a person from outside of the area but must involve people from within the audited area to be effective. An outside auditor will not fully understand the activities being performed in the area, nor will the audit results be well received by those who *do* work in the area. A mix of internal and external auditors works best.

Figure 6.3 Visible Shine standard used in hotels.

A consistent scoring system should be used for the audit, something simple that everyone understands—no fractions, decimal points, etc. The form provided in the appendix uses a scale of 0 to 100, much similar to elementary and secondary school grading systems. A score of 90 to 100 is excellent, 80 to 90 very good, 70 to 80 good, and so on.

It is important to formally recognize the results of the audit. Recognition can take many forms, and it is best to use a combination of methods over time. Most commonly, organizations recognize the best cumulative score and the most marked improvement. Awards can be in the form of modest gift certificates, a free lunch, or even something symbolic. Many companies have initiated a trophy as the form of recognition. The value of the award is less important than the recognition itself.

To Sustain workplace organization over time, it is critical to train new associates appropriately. Workplace organization, and Lean in general, can represent a whole new language to most people. Associates who are new to the organization must be taught this new language if they are to understand and participate in the 5S system. Unfortunately, in many organizations, little or no follow-up training is provided to new associates after the initial 5S effort. Training usually consists of 2 to 4 hours of reviewing 5S concepts, the terms, and the process in place to Sustain the work.

To help Sustain workplace organization, you can periodically schedule focused 5S kaizen events to improve the general flow through an area or to address a specific problem (e.g., the organization of a filing system, a particular quality problem, an issue with safety). The point here is that you need to use the 5S methodology periodically over time. This reinforces the idea that 5S is part of the Lean management system—the fabric of the organization and the way in which business will be conducted—and not simply a program with a beginning and an end.

The productivity benefits to improving workplace organization can be substantial, depending on the starting point of the organization. Workplace organization can raise productivity anywhere from 5% (if current conditions are fairly good) to 15% (if current conditions are quite bad). But remember, it is not all about productivity. Customer perceptions are even more important, and the benefits of reducing frustration and stress for people who work in an area are immeasurable.

At one company, the last 5S holdout in the office area was a woman in the accounting department. Her area was visibly disorganized, but she argued that it was organized the best way for her. The organization needed to demonstrate the amount of time and energy that she was losing each day due to the way she was currently organized. In turn, this would demonstrate to her that there was a better way to organize. She conceded to being observed for an entire day.

The observer noted the amount of time she spent searching for things and the distance that she had to travel in the course of her day, etc. In other words, the amount of time she lost performing non-value-added activities was tracked. The observer also noticed that the level of frustration and stress that she exhibited increased throughout the day. The woman frequently had to stay late to complete her duties, since approximately 20% of her day was spent on non-value-added activities associated with her lack of organization. When faced with the data, she was initially disbelieving, but ultimately conceded that the observer was right.

For the next 2 days, the two worked together to better organize her work area. This included organizing hardcopy and electronic files; setting up clearly identified in-boxes for the various information that people dropped off at her desk; better organizing in-process work using visual techniques to identify its status; creating basic standard work; and providing enhanced printing capability within her area. It was a stressful 2 days, as real changes were made. On the fourth day, the observer re-collected the data and found that 15% of the employee's time had been freed up because of the changes that were made.

The woman had not realized the degree to which the way she was previously organized affected her emotional state. With improved workplace organization, she no longer had to work long days. She no longer went home exhausted from the stress at work. "I actually have energy when I go home. Now I can get things done at home in the evenings," she said. She explained her initial resistance as, "I thought you were going to come in to my area and tell me to throw things out. I did not realize you were actually going to help me." This is a common misconception of 5S, one that is reinforced by many organizations that approach it as a "housekeeping" tool. It is also a common reason why 5S encounters sometimes strong resistance. The woman became a "poster child" for office 5S and its biggest proponent.

Mistake Proofing

Mistake Proofing requires people and organizations to find creative ways to eliminate the possibility that errors can happen. Mistake proofing goes well beyond the traditional approach to "quality assurance," which typically consists simply of inspection steps meant to catch defects. If defects persist, the common response is to add more inspection steps. However, the root causes of the defect go unaddressed, perpetuating an often-vicious cycle. Processes continue to spawn non-valued waste throughout, whether by creating the defect in the first place, looking for it (inspection is by definition non-value-adding), or correcting the problem (which should not have been created in the first place).

Too often, the cause of a defect is identified as "operator error." The corrective action involves communicating the problem to its originator and asking that it never happen again (until the next time that it does, of course). The assumption, in this case, is that the person alone has the *full* ability to control quality. This is in marked contrast to Lean Thinking, where the focus is on process. Think of it this way: can the process be designed in such a way to *prevent* errors from occurring? How can we make the process *less* dependent on the people performing the various tasks? In other words, how can we make the process more autonomous or self-managed?

Do not make the mistake of viewing this effort as a "dumbing down" of the process. This is not the correct view. It is about treating people with respect. Organizations *disrespect* people when they fail to provide the best processes and tools to their associates. It is disrespectful to waste people's time with processes that help create problems and the need to correct them. Nobody wants to complete a task all over again. It is frustrating and stressful to have to redo work. And, of course, the cost of failure in some situations can be huge to the customer and the organization, particularly in industries such as health care.

People do not intentionally create defects. Defects occur for a myriad of reasons, including rushing to meet deadlines (real or artificial), interruptions that cause a loss of concentration, inadequate training, and insufficient understanding of the needs of downstream processes. It is in the best interests of the people performing the tasks and the organization to find creative ways to mistake-proof processes. Mistake proofing is well documented within manufacturing. Luckily, different mistake proofing devices—specific techniques that prevent defects from being created—developed for manufacturing can be applied to office and service situations, even though the product is often less tangible. Consider these devices a launching point, intended to spark your own creative mistake proofing devices for your own individual applications.

Terms and Definitions

Before we discuss different mistake proofing devices, there are some terms that you need to be clear about. An "error" is the action that creates a "defect." Often, organizations install inspection steps in order to discover defects. Far too often, these inspection steps are incorporated toward the end of a process. This means that defects are discovered late in the process, often after more defects have been created. It makes more sense to identify an error before the resulting defect is created. This requires, at the very least, providing staff with the means to identify errors before they have been passed on as a defect. Even better still is to implement techniques that prevent errors from occurring by "mistake proofing," "error-proofing," or "poka-yoke" (all terms are synonymous). It is not always possible to achieve the mistake proofing ideal. However, with a little creativity, it is often possible to move closer and closer to it.

There are three levels of mistake proofing controls:

- Level 3: Discover the defect.
- Level 2: Identify the error.
- Level 1: Mistake proof.

Clearly, your goal is to implement Level 1 controls. But again, this is not always possible. Often, Level 2 controls are the best that are possible, and still represent an improvement over Level 3 controls. You also need to recognize that there are devices that fall somewhere between these three levels. Later, I will provide some examples of mistake proofing, and you will see how and why many fall short of ideal mistake proofing devices. In most cases, people can simply fail to use them. However, if they are implemented and used appropriately, they can help you prevent mistakes from being created and passed on and help you achieve Level 2. Your goal is to incrementally move closer and closer to the mistake proofing ideal.

Mistake Proofing Devices

In his landmark book, *Zero Quality Control: Source Inspection and the Poka-Yoke System* (Productivity Press, 1985), Shigeo Shingo identifies two types of mistake proofing systems:

- Control methods: When abnormalities occur, an operation is stopped.
- Warning methods: Warning methods call abnormalities to the attention of the worker. These are less powerful than control methods, as people can ignore the warning.

In addition, Shingo identified three general categories of mistake proofing detection methods:

1. The *contact method*: Methods in which sensing devices detect abnormalities by whether or not contact is made between an item and the sensing device.
2. The *fixed-value method*: Abnormalities are detected by checking for the specified number of motions in cases in which operations must be repeated a predetermined number of times. For example, instructions for assembly are provided in packets of 50. A check is made after 50 cases to see if there are any instructions left over that would indicate that there is a case or cases that did not have an instruction properly placed inside.
3. The *motion-step method*: Abnormalities are detected by checking for errors in standard motions in cases in which operations must be carried out with predetermined motions. For example, a label dispenser might have an electric eye that can see if the label was removed and attached to the item.

Each category of detection method can include many mistake proofing devices, and there are numerous "sensors" available. Although most have been created for manufacturing or production environments, with some creativity you can apply them in office and services, as well.

In the following paragraphs, I will provide numerous examples of mistake proofing devices, drawing on 12 of the more commonly used devices in manufacturing. Note that most are "warning method"-type systems rather than "control method" systems and as such fall short of the ideal Level 1 mistake proofing device. Nonetheless, each provides important benefits to the user. Also note that some could be considered as a different device, as there is some overlap between them. It is not important to categorize the device but to recognize the benefit of applying it to particular situations.

Mistake Proofing Devices and Examples

Guide/Reference/Interference Device

A solid piece of material that positions or orients something to guarantee its correct placement

Examples include recycling bins that have been designed with openings that allow only specific items to be inserted; therefore, only paper can be inserted in the paper bin, and bottles and cans in their correct bins. In health care, there are "sharps" medical waste containers, small red plastic containers used to dispose of used needles and any sharp objects related to medical procedures. The container is sized to prevent other objects from being placed within it, preventing different categories of waste from being mixed together.

Template/Checklist Devices

A pattern used to represent an accurate copy of an object or to guarantee accurate position
A list of items that must be present in order to guarantee quality

Templates and checklists are probably the most commonly used devices in office and service environments. Although they may fall short of pure mistake proofing (because people can fail to use the template or checklist), they certainly move in the correct direction by detecting errors before they become defects. Checklists are used in almost all service environments, including health care (for surgical instruments and supplies used during procedures) and air travel (pilot checklists).

Light Contact Electrical Devices

Devices that use light contact to confirm the presence, position, dimension, etc., without damaging or otherwise altering the item being detected

In many ways this is what required fields in a computer program represent. Users are notified immediately when required information is not present. Perhaps, the user cannot proceed further unless the required information is input. This represents a "control method." Bar codes are also such a device. Bar code readers can detect the presence of a specific item, avoiding the need to manually enter the information (a method that is much more prone to error).

In health care, a patient's identification bracelet can be scanned by a doctor and all of the patient's information transferred to a handheld device. The doctor is then able to review and place their orders. If, for some reason, an order is contradictory (e.g., a potentially dangerous prescription drug interaction), the device will immediately notify the physician, preventing medical mistakes.

Counter Devices

> A mechanical or electrical device that keeps track of numbers. This device is intended to negate the need for people to track numbers, and thereby eliminate the possibility of errors in doing so.

What does this device look like in practice? In the legal profession, attorneys use time recorders to ensure accurate billing. Another example is order entry systems that count the number of items in an order, which can help the order entry associate verify that the correct number of items have been input. This approach falls short of the ideal, since it is still possible that the incorrect item has been entered, but it is a start. Similarly, packing and shipping systems can also count the number of items packed (often through scanning) and compare them to the number of items on the order.

Odd-Part-Out Devices

> This is a form of counting that does not rely on a counting device, but rather visual techniques. The intent of this device is to detect when something is missing or out of place.

This device can take very simple forms; for example, by drawing a continuous diagonal line on the spine of books stored on a shelf. A broken line indicates that a book is missing or has been placed in the incorrect location. Medical records often use a visual coding system that makes it very clear if a particular record is missing or is out of place. Another health care example involves using organizers to detect that an item is missing; for instance, a sponge left in a patient during surgery. Job folder organizers that include a place for all required items are yet another example. Again, these folder organizers make it apparent that a required item is missing. This device can also take the form of required fields in a computer system with the computer highlighting fields that have been left blank.

Think back to our "Set-in-Order" standard—the second "S" of the 5S system. Set-in-order is often described as "a place for everything and everything in its place." When every item has a home address, it is readily apparent when an item is missing. These techniques are very helpful in a wide variety of service organizations. For example, service technicians with highly organized and labeled kits for tools and materials can identify missing items *before* they are needed and prevent the error of arriving at a customer site without all required items.

Sequence Restriction Devices

> Only the desired sequence is possible. This is intended to ensure that the associate performs a given series of steps in the correct order.

For example, therapeutic treatment machines used in health care require a specific sequence of buttons to be pushed before treatment can be delivered to prevent accidentally turning the machines on. In addition, many computer screens have been designed to allow only a particular sequence of data entry.

Standardize and Solve Devices

> When weight, dimension, or shape standards are in place, devices can often be developed to identify nonstandard conditions. These devices are intended to easily recognize when a clearly defined standard has not been met.

For example, self-checkout lines use scales to confirm the weights of items to ensure more accurate processing. And we have all had to re-enter our e-mail addresses as a comparison to one we have already entered. If the two entries do not match, then we have to re-enter ("solve") the information. Now, it is possible that you could wrongly enter the address twice, but it is highly unlikely.

Here is a common example. Let us say that an input to a computer screen does not meet some expectation, such as the number of digits in a phone number, a credit card, etc. The system will automatically require you to re-input the information. Other systems have been designed to detect the presence of particular digits or sequence of digits. Again, this may fall short of an ideal mistake proofing device (i.e., people can still input an incorrect phone number, which would still have the correct number of digits), but it certainly is a step in the right direction.

Critical Condition Indicator Devices

This device detects two types of conditions:

- The presence or absence of a specific, visible, preset quantity, weight, volume, etc.
- Fluctuations in a nonvisible condition, such as pressure, temperature, current, fluid flow, etc.

There is a fair amount of overlap between this and the previous device, "Standardize and Solve." In fact, many of the examples to come can easily be considered that type of device. In an office and service environment, "time" is often the critical condition, and there are a number of ways to detect whether work is being conducted within the appropriate timeframe.

For example, folders color-coded by day can indicate the time that work has moved through a particular area or areas. The system can clearly identify work that has not been processed within the desired timeframe (see Figure 6.4). In this example, the goal is to have all orders processed through this area within 24 hours. Each day has an assigned color. A colored tab is attached to the folder that corresponds to the day that the order was received. This would provide an immediate warning that a particular order was not processed in a timely manner.

Another example turns up in health care, where prescriptions electronically submitted by doctors are cross-checked against databases to prevent errors such as undesirable drug interactions. There are also numerous examples of these devices on office equipment such as copiers and printers (e.g., "toner low" warnings).

Delivery Detection Devices

A passageway where detection devices can be used

One example of a delivery detection device is the use of radio frequency identification devices (RFID) to identify when an item has been removed

Figure 6.4 Colored tabs indicate number of days in an area.

from a storage area. RFID tags are often used in retail operations to help maintain inventory accuracy. Have you ever been stopped by an alarm as you are leaving a store? That is another example. A security tag on a product that detects when unpaid items are leaving a store is the culprit in that instance. In both cases, there must be clear "passageways" for such devices to be effective.

Stopper/Gate

This device is a solid piece of material that guarantees that a certain operation is *not* performed. The stopper/gate is intended to prevent a step from being performed.

Perhaps only certain fields are to be completed, depending on the order type. Programs are available in which certain fields cannot be filled, and/or the system will skip to the applicable fields. Alternately, a pop-up on a computer screen might ask the person inputting information to verify that what they entered is correct. At one company, which offered similar products for use in the United States and internationally, customers using an online system often ordered the wrong voltage for the product. The company implemented a pop-up message asking the customer where they intended to use the product to better ensure that he or she ordered the correct item.

Sensor Devices

An electrical device that detects and responds to fluctuations in characteristics. There are numerous types of sensors, including beams, vibrations, displacement, positioning, and area sensors.

Some of you may work in organizations in which you need to insert your identification badge into a computer before it can be used, in order to maintain security protocols. That is one example. The RFID and security tag examples we explored previously could easily be included under this device, as well.

Mistake Proof Your Mistake Proofing Device

Where possible, you should use a combination of mistake proofing devices. Just as you back up your computer files, you need to back up your mistake proofing device.

When you use more than one device, the probability that you will create an error further decreases.

There are some techniques that do not neatly fall into the device categories provided, but are nonetheless very pertinent to the office and service environment. For example, reducing the need to manually transcribe or transfer information can greatly reduce the opportunity for errors to arise. Reducing writing can work toward mistake proofing by reducing legibility issues through the use of check boxes, electronic prescriptions, etc.

Merging your information sources into a single source is another example of mistake proofing, as information that resides in different locations will more likely become incongruent over time. Again, reducing the opportunity for errors to arise is mistake proofing itself.

Setup Reduction or Quick Changeover

You may be surprised to learn that even techniques developed mainly to facilitate the rapid changeover of manufacturing equipment (i.e., setup reduction, quick changeover, or "single minute exchange of die" (SMED) concepts) can be applied effectively to office and service environments. Let us define a changeover as the process to switch from one "type" to another. A type can be a material (e.g., different paper sizes in a copier), time (e.g., from one reporting period to another), or a person (e.g., from one customer to the next).

It is important that you recognize when particular activities warrant a changeover or setup. There are numerous changeover or setup processes in office and service environments. When you close accounting records for 1 month before moving on to the next, that is a changeover process—and it is common to most companies, regardless of industry. In health care, examples include changing over operating rooms from one procedure to the next and treatment rooms from one patient to the next. The hiring of one person to replace another can qualify as a changeover, and implementing a formal changeover process can help make the transition from the departing employee to the new employee as quick and effective as possible.

When you implement setup reduction or quick changeover, you need to distinguish between "internal" and "external" activities. For a changeover process, let us define internal activities as the activities required to process the next "type" that can only be performed when the current process stops. External activities are activities that can be performed either before or after the internal activities, and without the need to stop the current process (see Figure 6.5).

External activities often are preparatory in nature. Let us take a look at a typical hiring process as an example. What activities can you perform before the existing employee leaves? First, you might set up a user account on the computer system for the new employee before he or she even arrives. If circumstances allow, you might be able to train the new employee before the existing employee leaves, ensuring continuity in the role and process. Continually maintaining a file of qualified applicants for each given position is another example of an activity that can be performed "externally."

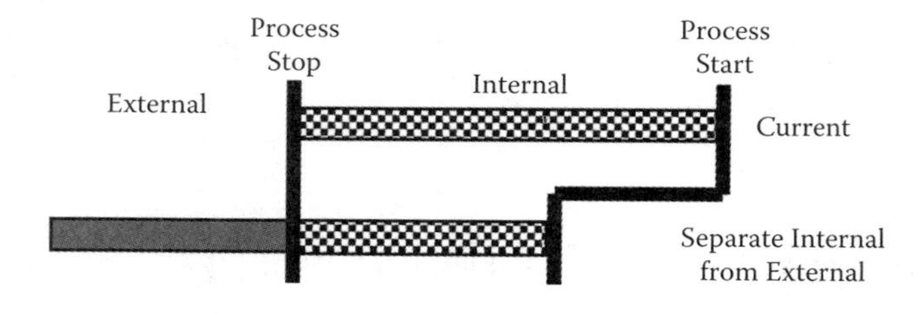

Figure 6.5 Distinguishing internal from external activities.

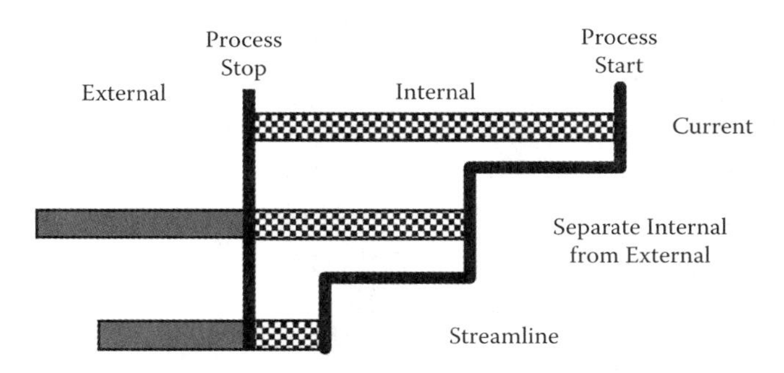

Figure 6.6 Results after streamlining.

Another key concept of setup reduction or quick changeover is to streamline all activities, internal and external (see Figure 6.6).

In streamlining, you identify opportunities to reduce the time needed to complete all activities. Let us continue with our hiring process example. Think back to the teaching techniques we discussed in Chapter 2, under Job Instruction. Applying these techniques can significantly reduce the time required to train a new employee, whether it is done before or after the existing employee departs.

We touched briefly on month-end closing as an interesting example of the application of quick changeover concepts. Let us look at that in a bit more detail. In a month-end closing, a company will "close the books" for one accounting period, and start anew for the next period. Often, this process requires a great deal of effort on the part of many people in the organization, both accounting and operations personnel. This effort occurs immediately after the end of each month, as individuals and the company as a whole work hard to complete the process. After all, managers want to know the results for the period just completed, and they need the reports generated from the process to do so. Accounting and other business services personnel work quickly to complete this process, since in most cases, the longer it takes, the more disruptive it can be to the operation. In spite of this sense of urgency, the process still takes up to 10 business days to complete in most companies.

However, there are many month-end closing activities that can be performed "externally." In other words, as much as 50% of these activities can be completed *before* the end of the month. Take the time to identify which activities can be completed ahead of time, and you can reduce the time to close to less than 5 days, and in some cases as quickly as 1 day.

Let us provide one more example of setup reduction from health care; this time, in a physician's office, with you as the patient. In most physicians' offices, you will arrive at an appointed time and wait for a doctor to become available to see you. The receptionist will provide required paperwork for you to complete while waiting. At some point in time, you will be led back to an examination room where typically (you guessed it) you will wait some more. Does the overall wait time seem excessive to you? It sure does to me.

But what if your wait time could be reduced? Would you be more satisfied as a result? What "external" activities could be completed *before* arrival to the office? For example, what if you could complete the required documentation prior to your arrival? You would still need to do the paperwork, but you could schedule it for a time and place that is convenient for you. Your overall time in the physician's office is reduced, and I can almost guarantee that as a patient you will be more satisfied as a result.

Reading through these examples should help you recognize the opportunities that exist to apply setup reduction or quick changeover concepts within your organization. Obviously, how you will do this will vary based on the nature of your processes—but taking the up-front time to do this work can result in vast improvements in efficiency—and your stress level!

Chapter 7

Functional Applications of Lean

As I have emphasized in earlier chapters, application of key Lean concepts must always take place within the context of overall value streams. Remember, most value streams involve multiple departments. However, we do need to recognize that most companies are still not organized by value stream, and that most people still think functionally or departmentally. If this sounds similar to your organization, do not despair, because in this chapter, we will review common functions and how to apply Lean to them. Obviously, we cannot cover all of the possible functions across the wide variety of existing companies and industries. Instead, we will stick to the most common functions, including

- Sales and Marketing
- Purchasing
- Accounting
- Customer Service
- Human Resources

Our focus will be to reduce the time and effort required to perform the various activities required within each function. But you will also need to ask yourself what you will do with your newly available capacity. To get there, you need to ask yourself another question: "When is the [insert function here] department truly adding value to the internal or external customer?"

In other words, what activities should be performed by these functions to ensure the organization's continued success? These are the activities that people should perform with their newly available capacity.

I cannot emphasize this enough: you need to make every effort to resist the path of headcount reductions. Employees will no longer support the key Lean concept of continuous improvement if they are concerned about job security. Of course, attrition—people leaving on their own and not being replaced—is acceptable. But, for those who do remain, you need to leverage the available capacity in ways that create *more* value. In this chapter, I will show you how.

Remember the general approach to implementing Lean that we covered in the Introduction? Again, the four steps are stabilize, standardize, visualize, and improve. The actual approach that you will take will depend on the existing state of your function or department. This is no different from applying Lean to an entire value stream involving multiple departments. If a current process is unstable or does not provide predictable and acceptable outcomes, then that is where your Lean effort must start. If the process is sufficiently stable, as is most often the case, then standardization is the starting point. We will review each common function within the context of these four steps.

Sales and Marketing

In all typical Lean implementations, there comes a time when the focus (and the pressure to change) moves to the Sales and Marketing functions. Lean efforts in operations and support areas will free up capacity, as waste is eliminated and processes streamlined. Invariably, your next question will become "What do we do with the available capacity?" To answer this question, Sales and Marketing personnel must identify opportunities within existing and new markets. They need to use *existing* sales and marketing resources in smarter ways, by applying Lean Thinking. Unfortunately, there is strong resistance to applying these commonsense principles to Sales and Marketing, often exceeding that encountered with other functions.

Once again, the most common arguments center around the variable nature of, in particular, the sales process. Sales professionals often take pride in the fact that "no customer is the same," "no sales situation is alike," etc. They often view the selling activity as "creative," and of course, argue that "Lean does not apply to creative processes." You will also often find that salespeople are independent in nature; in many cases, they are always on the road and return to the office only for periodic meetings. This can result in the loss of a sense of belonging with co-workers, as well as a very real disconnect from important business processes. And unfortunately, many sales personnel actually take pleasure and pride in the independent nature of their position, creating additional obstacles to overcome as your organization applies Lean.

You will find similar arguments in marketing processes, but to a much lesser extent. Again, marketing is highly creative but, fortunately, most marketing professionals will recognize that the activities that they most often engage in are indeed processes to identify information needs, sources of information, and the means to analyze the information. What is often missing is ongoing *management* of the marketing *process*, as there tends to be a project focus to this function, with marketing campaigns that by definition have a beginning and an end.

You need to overcome the general lack of process focus if you are going to successfully apply Lean concepts to the Sales and Marketing functions. But—and this is an important distinction—you do not actually apply Lean to functions, but rather to the *activities* that those functions perform. So, first, you need to identify the activities that are regularly performed in these functions.

Stability Issues with Sales and Marketing

Instability is often caused by a complete lack of process definition. For example, sales personnel are often left on their own to determine the process that works best for them in their regions of responsibility. By now, you will have realized that this has consequences for the rest of the organization, such as inconsistent order-related information across regions, extended learning curves for people assuming responsibility for a region, etc. Therefore, defining the process is often your first key step.

There are numerous sales models, all of which involve some variation of the following:

1. Identify new opportunities (typically through marketing efforts).
2. Follow up on new opportunities and establish relationships (e.g., initial contact).
3. Identify customer needs (e.g., problems they are trying to solve) and buying parameters (e.g., budget, timing).

4. Identify customer decision-making process (e.g., who, how, when).
5. Obtain order.
6. Do postsale follow-up.

Each step represents a process that can be further defined, standardized, managed (even visually), and improved upon. Each organization should have a process that they can readily describe, and which all sales personnel can commit to and reference as needed.

In practice, most sales personnel are following something similar to this process, or at least are following particular steps, though they may not have thought of it in these terms. People will often overlook important steps simply because of a lack of process clarity or other distractions. For example, many salespeople overlook the postsale step, thereby missing out on the chance to identify future selling opportunities.

Very often, you can mitigate these issues simply by putting on paper existing, poorly defined process steps, agreeing on terms and definitions, etc. You will see how important this is shortly, as we move on to "Standardize" and "Visualize." But for now, know that once a selling process is in place, your organization can achieve much-needed stability.

For marketing processes, effectiveness needs to be considered first and foremost. Often, organizations are not aware of the effectiveness of various marketing approaches and processes (e.g., direct mail, promotions), and whether and how they affect the desired results. You will need to gather some data to determine which approaches are more effective, and can then standardize them accordingly.

Standardizing Sales and Marketing Processes

It might help to think of the sales process as the "what to do"—the steps that each salesperson will follow. When you standardize the sales process, though, you need to think in terms of the "how"—the manner in which the steps will be performed. Again, standard work is a foundational concept of Lean. Recall from Chapter 2 that it includes the "key points" that define the "how" and the "why" to perform particular steps. Key points for any process typically involve quality, efficiency, the time necessary to complete a step, and possibly the timing of a step's performance. For example, a given step must be performed in a particular way to ensure an acceptable quality result. In sales, this might mean that sales personnel must fully complete order forms so that an order can be processed accurately, and that order forms must be submitted at particular times so that they can be processed in a timely manner.

Defining the "how" and the "why" will better ensure that sales personnel perform their activities in a way that meets the needs of "internal" customers (e.g., in-house order processors, scheduling personnel, shipping). In their rush to bring in orders, sales personnel often overlook these needs, causing more non-value-added processing waste—even for the salesperson himself, who has to provide additional information at a later point. It is always better to do it right the first time, in sales or any other function.

We need to distinguish between standard work and "style." Sales personnel often resist the concept of standard work because they confuse the two. Establishing standard work does *not* mean that each salesperson must practice the same sales style. One salesperson may establish rapport to build relationships by discussing subjects of personal interest, while another may choose to discuss recent business news. Both can be effective, and both can still follow a standard selling process.

Further, standard work can be adjusted based on true customer needs. At one company, the process time to be spent on a telephone call taking new orders was intentionally varied, based on cultural differences between customers from the United States versus Europe. Customer surveys clearly indicated that the customers from Europe preferred to spend more time on the telephone, while customers from the United States preferred faster phone order times. The differences were taken into account in the standard work documentation. Following the established standard work meant that order processors, even newly hired ones, could ensure a positive experience over time for customers from different regions of the world.

SIDEBAR Tailoring Standard Work

One company did a lot of business with school districts in the United States. But all school districts are not created equal, and this organization needed to create different standard work for several categories of school districts. For example, there were very real differences when dealing with school districts in the state of New Jersey, where there are over 1,000 districts—essentially one for each town—and Florida, which has less than 10 covering the entire state. In New Jersey, a much more personal approach was necessary, with additional time required to develop relationships with those involved in the decision-making process to purchase goods and services. In Florida, procurement personnel were more "professional," more interested in completing the transaction than developing relationships.

In addition, standard work needs to be applied to all of the "secondary" activities in which sales personnel are involved. What are secondary activities? They can range from travel planning to expense reporting to generating sales reports and more. When organizations implement Lean, they often overlook the potential to streamline these types of activities, but it is a tremendous opportunity to reduce non-value-added processing waste. On average, you can free up 10%–15% of a salesperson's time by streamlining "secondary" activities. And since time really *is* money, this extra capacity can be leveraged into proactive sales generation activities, allowing some organizations to realize a proportional 12%–15% growth in sales from the previous year. These gains required a series of "cubicle-level" kaizen events to identify the secondary activities, prioritize them, study them in detail, and determine ways in which to streamline them.

One way to standardize marketing activities involves the identification of sales opportunities. Companies will often purchase databases that contain market information that can be used to identify opportunities; for example, a list of companies that have purchased similar goods and services in the past year. These databases change over time, as they are updated with more recent information. Unfortunately, it is often left to the different users to determine the best way to extract or mine the data. The result is that different people use different criteria for identifying the opportunities, and what actually works best often goes undefined.

At one company, a cross-functional team of people identified the key criteria that would be used to extract the desired data. These criteria were developed based on empirical data and the collective experience of the team. Historical data revealed that projects that shared certain characteristics (e.g., value, type) were much more likely to result in future sales. This work led to standardized criteria and keywords for searching and filtering the databases, which in turn led to more focused and thoughtful sales efforts.

Now that you have standardized your sales and marketing functions, let us make them visible.

Making the Sales and Marketing Function Visible

As you will recall from earlier chapters, visual management is another key Lean concept. In most organizations, time is at a premium, and visual communication is an effective and efficient way to make visible the key aspects of important processes. First, let us take a look at some common aspects of the sales function:

- The activities that a salesperson *should* be engaged in (e.g., prospecting, following up on new opportunities, following up on existing customers)

- Nonstandard conditions (e.g., *not* completing required activities when they were required, information quality issues that arise, spending more or less time than a sales associate should on a particular activity)
- Performance (e.g., meeting personal or team goals, processing new opportunities within a desired timeframe)
- Queues of work (e.g., opportunities that need to be attended to by particular sales associates, quotes and orders that need to be processed in a timely manner)

So what are the organizational benefits of making these aspects of the sales function visible? First, the sales and marketing department could have real-time, or near-real-time, access to information critical to its performance. Too often, sales organizations wait until the end of the fiscal period, when sales reports are generated to identify performance-related problems. But a Lean enterprise has a very short management timeframe, and you need to review performance frequently, in a way that does not require a lot of non-value-added effort (such as generating time-consuming reports). The process needs to be visible. Period. Today, many computer-based sales systems provide workflow capability or ways to track the queue of sales opportunities at various stages of the process. Although these systems can provide much-needed visibility, you can also use simpler methods. For example, for a sales associate role, you can develop standard work that defines the activities he or she is to perform, the time and/or timing of the activities (e.g., 2:00–3:00 p.m. each day), the number of tasks to be completed (e.g., five cold calls per day), and some visible evidence that the activity was completed. Office personnel can manually track problems with information quality as they encounter them. Often, all you need for this is a simple dry-erase board or boards in the sales office (see Figure 7.1). Of course, when you involve "outside" sales—sales personnel who work out of their home or are constantly on the road—electronic techniques tend to work better. But do not overlook simple, visible, worker-managed techniques. When used correctly, they can pay large dividends.

Now that you have standardized your processes and made them visible, let us shift our focus to continuous improvement.

Improving the Sales and Marketing Function

Applying Lean concepts to the sales function, in particular, standard work and visual management, can provide important benefits to the organization.

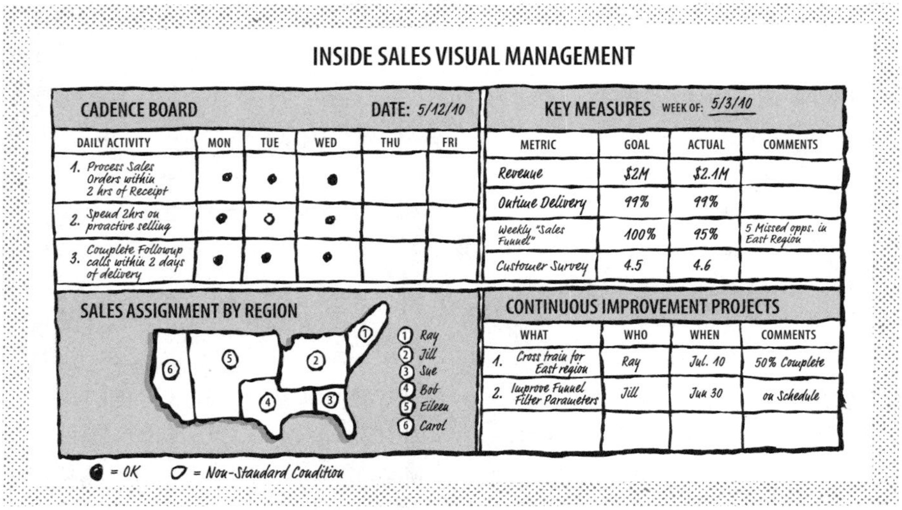

Figure 7.1 Inside sales visual management example.

By eliminating waste from various sales activities, you can free up valuable capacity, which can then be used to proactively generate sales. This *increases* the value that an organization delivers to its markets, which is the true goal of every Lean enterprise.

Sales managers, and managers in general, need to spend a significant portion of their time on improving sales function and performance. Unfortunately, most sales managers find themselves caught up in non-value-added activities that take time away from this critical activity. But we are in luck, because using our newly implemented visual cues, sales managers can more easily identify opportunities for improving both overall processes and individual performance.

Let us say that a sales associate is following established standard work, but still not meeting performance expectations. Since the associate is already working with standard conditions, it should not take the sales manager too much time to determine root causes of the problem. Maybe it is an issue with "style" or the manner in which the associate interacts with potential customers. Or maybe another associate is consistently submitting incomplete or inaccurate information. In a Lean organization, this shortcoming will become visible fairly quickly, since Lean companies review performance often and identify problems as they occur. And when problems become evident quickly, sales managers can work with their associates to resolve them equally quickly.

Next, let us take a look at the mix of activities that sales associates are expected to perform. Most people (including you and I, I would wager) put off activities that they do not want to perform. A sales associate may avoid making cold calls because it is unpleasant; after all, it is much easier to call existing customers, with whom the associate already has a relationship. But important opportunities may be lost by avoiding cold calling. By using visual techniques to tune in to this type of issue, sales managers can work to address them in timely ways.

Dedicating the necessary resources (e.g., the time needed to conduct kaizen events) to implementing Lean in the sales and marketing function can be challenging. Again, to make it work, you need to involve your employees in the improvement process. Involvement can take the form of collecting data to identify improvement opportunities, developing and agreeing on standard work, and participating on a kaizen team to resolve specific problems. The benefits to the sales team will far exceed your initial investment in time and resources.

Purchasing

The results that you can achieve with Lean will be limited by the performance of your suppliers. In almost every Lean effort, there comes a time to move the focus to the supplier base, which in turn requires the involvement of your Purchasing department. Purchasing's basic function is to obtain goods and services from outside sources to fulfill a need that cannot be met with existing internal capabilities. Unfortunately, Purchasing does not always completely understand the needs of the person who requested the good or service, but is nonetheless responsible for communicating with suppliers. This can result in quality issues with the purchasing process, and the purchasing of goods or services that are not needed. There is more to purchasing than meeting the specification at the lowest possible price, but unfortunately, that is how many organizations view this important function.

Purchasing personnel need to search for *complete* solutions to meet all of the needs of the organization. These include consistent quality and service performance, which go beyond price. What do consistent quality and service look like? When purchasing goods, these include the frequency of delivery from the supplier, supplier-mandated purchase quantities, supplier-provided container amounts, quality, reliability of delivery, terms of payment, and the like, all of which will affect the organization's ability to improve its performance. When purchasing services, these include the quality and timeliness of the service, as well as the ability of the service provider to work with members of the organization and to follow applicable standard work.

Purchasing staff need to collaborate with other members of the organization—their internal customers, including operations and accounting personnel. These existing relationships might be strained and in need of repair; often, operations and accounting associates will complain that purchasing personnel do not entirely understand whom they work for. But let us recall Chapter 1, where we discussed value stream management and, specifically, the "requisition-to-pay" value stream that included the Purchasing, Operations, and Accounting departments. I cannot stress this enough: the value stream can only be improved through collaboration among these functions.

To implement Lean in the purchasing function, you need to develop long-term supplier partnerships. There is a degree of give and take in these partnerships; for example, if you assist the supplier in solving problems related to its product or value stream, both you and the supplier will benefit. Maybe your Purchasing department could facilitate a kaizen event, on a quality issue, at your supplier's location. You will likely face some resistance from Purchasing at first, since associates are so busy with the "transactional" aspects of the function (e.g., processing purchase orders, tracking down receipts, expediting) that there is little to no time for partnership- or relationship-building activities. This is precisely why the initial focus of Lean efforts in Purchasing is to streamline transactional activities, freeing up time, and making room for changes to the Purchasing role.

Stability Issues Relating to the Purchasing Function

Most organizations have well-defined purchasing processes, but that does not mean there are no issues with the predictability and consistency of the output from these processes. Issues that create instability usually involve the *existing* suppliers' performance. A great amount of time and energy is lost dealing with a few poorly performing suppliers, but it is not always possible to find alternative suppliers. Your only remaining option then becomes to assist the existing supplier in addressing its own issues. Again, you will likely meet with some internal resistance to this, largely because of a lack of available time. But if you make the investment of time, you will be rewarded with much-needed stability.

Why would you accept poor supplier performance? Besides a lack of alternatives, another possible cause of a continued acceptance of poor supplier performance is the metrics to which people involved in the purchasing function are being held accountable. As I hinted earlier, many purchasing functions are focused on price, and price *only*, not the full cost of the existing relationship. A relatively simple change to the measurement system, from a system based solely on price to one based on all aspects of supplier performance, can suddenly trigger Purchasing personnel to seek and find alternative solutions. Measurements drive people's behavior, and the Purchasing function is no different.

Another possible cause of instability is changing priorities and due dates for purchased items. This most often occurs in organizations that depend on complex computer systems, such as Management Resource Planning (MRP) systems, to identify and plan items to be purchased and manufactured. These systems generate numerous reports to change due dates for specific items (e.g., "move-up," "move out" signals) as they perform the (often overly complex) scheduling function within the computer system. Purchasing personnel are then compelled to respond to the system's suggestions for fear of overlooking something. Unfortunately, there is a lot of "noise" in these systems, as they are very sensitive to minor changes in the data upon which they rely. An insignificant change in the schedule for a manufactured item can create a suggestion by the system to change the schedule for a purchased item, even when it is not really necessary. Small changes in one part of the system can have a compounding effect throughout and, in these cases, you need to develop alternate and simpler processes to identify, plan, and trigger ordering of purchased items. These processes can vary, and *all* can be standardized.

Standardizing Purchasing Processes

Do not misunderstand what standardization means in Purchasing. It does *not* mean that all goods and services must be purchased in the same way. Unfortunately, the standard process in many organizations is to initiate a requisition form for all purchased items, which in turn generates a purchase order. These practices usually are established under the guise of standardization and control. But organizations can and should implement multiple processes for purchasing goods and services. This is because the nature of goods and services can greatly vary, as can the needs of the requestors (the internal customers) and the abilities of the suppliers. You need to standardize each process, once it has been proved to be effective and efficient.

If your company purchases goods, materials, and supplies, you can establish simple supermarket pull systems (see Chapter 4). You can also implement supplier-managed material processes. In supplier-managed systems, the supplier is responsible for monitoring and managing the inventory. Another way to purchase items is by use of credit cards that have been issued to specific members of the organization. But regardless of the process, standard work can be developed that contains appropriate controls. The key is to keep each process as simple as possible. The result is that *less time* will be required from Purchasing personnel to perform transactional activities. This creates more time for the partnering activities previously described, or strategic activities, such as developing alternative sources.

Making the Purchasing Function Visual

Remember the examples we reviewed in Chapter 5? All of these can be applied to the Purchasing function. It is critically important to monitor that the department is keeping up with the demand, so monitoring Purchasing's various queues of work must be part of the visual management system. This visual management system will need to include a way to monitor responsiveness and the timely processing of work, as well as some means of tracking quality performance of the department. You will also need to monitor the performance of the supplier base (see Figure 7.2). And, as with all other functions, you will

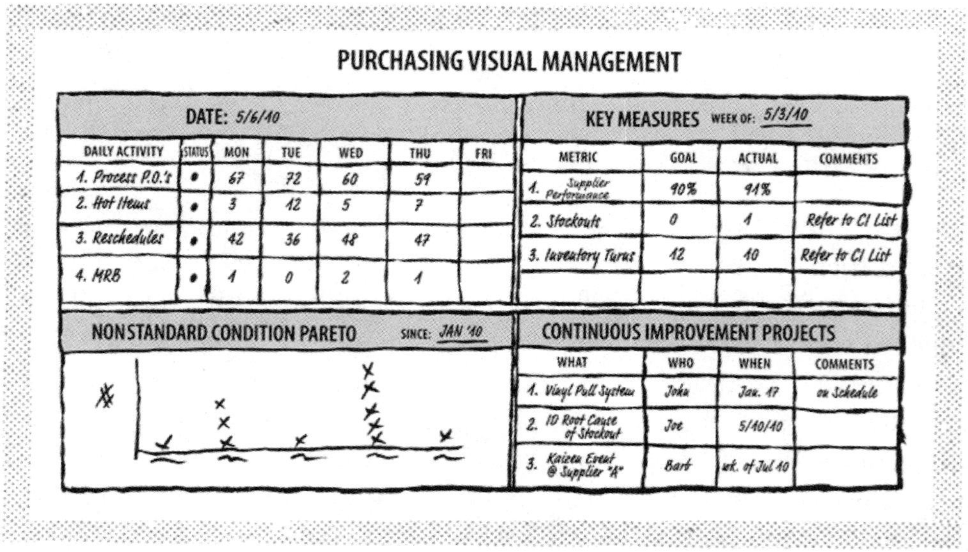

PURCHASING VISUAL MANAGEMENT

DATE: 5/6/10

DAILY ACTIVITY	STATUS	MON	TUE	WED	THU	FRI
1. Process P.O.'s	•	67	72	60	59	
2. Hot Items	•	3	12	5	7	
3. Reschedules	•	42	36	48	47	
4. MRB	•	1	0	2	1	

KEY MEASURES WEEK OF: 5/3/10

METRIC	GOAL	ACTUAL	COMMENTS
1. Supplier Performance	90%	91%	
2. Stockouts	0	1	Refer to CI List
3. Inventory Turns	12	10	Refer to CI List

NONSTANDARD CONDITION PARETO SINCE: JAN '10

CONTINUOUS IMPROVEMENT PROJECTS

WHAT	WHO	WHEN	COMMENTS
1. Vinyl Pull System	John	Jan. 17	on Schedule
2. ID Root Cause of Stockout	Joe	5/10/10	
3. Kaizen Event @ Supplier "A"	Bart	wk. of Jul 10	

Figure 7.2 Visual management techniques applied to purchasing.

need to post standard work for the various activities that are performed by department personnel. As with all functions, the visual management system should drive the continuous improvement efforts of Purchasing personnel.

Improving the Purchasing Function

When improving the purchasing function, your main objective is to create improvement in the overall performance of the supplier base. Performance of the supplier base includes total acquisition cost, service, and quality. Total acquisition cost includes elements other than price, such as inventory carrying cost. Purchasing must be part of the ongoing effort to decrease the amount of inventory stocked while maintaining or even improving service to both internal and external customers. To do this, suppliers and Purchasing personnel alike will need to participate in kaizen events conducted in the company, or at supplier sites.

For companies that periodically develop new products, Purchasing needs to participate in the improvement of the development process and, further, they (and appropriate suppliers) *must* become involved earlier in the development process. In most organizations, this type of involvement typically comes much too late, if at all. This sometimes happens because Purchasing personnel do not have the time available, and so get involved later, after design details have been worked out. The belief, although incorrect, is that this will be a time-saver in the long run. Additionally, current development processes might not trigger Purchasing and supplier involvement until later—because that is "how they always did it." But whatever the reason, an investment of time by Purchasing earlier in the development process will provide tremendous benefits to everyone.

Responsiveness to internal customers, operations and accounting in particular, will often be a focus of improvement. Cross-training in the established standard work will often pay great dividends, providing much-needed flexibility within the department.

Purchasing personnel can significantly contribute to many improvement opportunities in the organization—but only if you first free up time so that they can participate more fully than they have in the past.

Accounting

All organizations, manufacturing and services alike, have an accounting function. But all too often, accounting professionals sit on the sidelines during Lean implementation. This most often happens because of the inaccurate perception of Lean as strictly applicable to operations. However, most accounting processes are interconnected with operational processes and, often, obstacles to the implementation of Lean concepts in operations lie in existing Accounting practices and processes. Arguments against change often blame the Accounting department, which makes it even more critical to involve accounting personnel in Lean efforts in operations. Further, people who work in accounting have skills—specifically, analytical skills—that can greatly aid improvement efforts. These skills can be very useful in the effort to collect and understand data, assess alternative solutions, and measure the impact of changes that have been made.

You will hear the same argument from Accounting that you will from the other functions: "We do not have time to participate in improvement efforts." Again, your early focus in Accounting is to streamline current accounting processes, thereby freeing up time that can be devoted to business improvement. The accounting processes include setup of new customers, invoicing, collections, paying suppliers, month-end close, budgeting, inventory valuation, financial reporting, auditing, payroll, and other activities.

Stability Issues in the Accounting Function

Accounting processes are often very well defined and quite stable, simply because of the nature of the role. Consistency is one of the four tenets of Generally Accepted Accounting Practices (GAAP) (the others are Materiality, Conservatism, and Matching). However, there may be instability in the reporting of financial and other performance measures that Accounting generates. Particular measures may not appear correct to Accounting, triggering a lot of non-value-adding activities to investigate. In these situations, you need to identify and address the root causes of the system issues that generate the questionable data. These system issues are often related to the operational processes that are the sources of the data. We will discuss this in more detail when we explore standardization, which is typically where the Lean effort in accounting begins.

Standardizing Accounting Processes

Most accounting processes, including their timing and scheduling, are already well defined and usually adhered to by members of the organization. Accounting departments in most companies closely follow a calendar that defines when activities are to be performed. However, that is not to say that personnel in Accounting follow standard procedures as closely as they should or that all key points have been clearly identified. Nevertheless, most accounting firms and departments already have implemented standard work to a large degree.

But current accounting processes can be inefficient, if not for Accounting personnel, then for others in the organization. You need to design your accounting processes in ways that are easy for *everyone* to follow. Accounting personnel often express frustration with others who do not follow existing procedures or do not complete tasks on time. When Accounting personnel have to chase after people to get the information they need, you need to work on simplifying the process for everyone. Remember, a key concept of standard work is to *streamline* the work. There is a much better chance that people will follow standard work for an activity if it requires 5 minutes to complete rather than the current 10 minutes.

Thankfully, you will find plenty of opportunities to streamline existing accounting processes. Accounting is a unique function in that the industry standards include GAAP, which accounting professionals often point to as a reason not to change. The four tenets of GAAP are Consistency, Materiality, Conservatism, and Matching, all of which are actually very compatible with Lean thinking. How so? GAAP defines Consistency as "treating like transactions in the same way in consecutive periods so that financial statements will be more comparable than otherwise." Well, the purpose of standard work, a foundational concept of Lean, is to achieve *consistency*. GAAP defines Materiality as "when the judgment of a reasonable person relying on particular accounting information would have been changed or influenced by its omission." And Lean thinkers always try to distinguish between the significant (i.e., material) and insignificant. GAAP defines Conservatism as "a reporting objective that calls for the anticipation of all losses and expenses, but defers recognition of gains and profits." In practice, Lean thinkers are quite conservative–methodical in their approach, fact-driven, and they use data to support the design of various elements of the Lean management system. Lean thinkers rarely take any true risks. Finally, GAAP defines Matching as "the concept of recognizing cost (expenses) in the same period when the related revenues are recognized." Flow is a key concept of Lean. And the objective of flow is to reduce the overall time to process materials and information, and to deliver services. This is fully supportive of the objective of matching, or matching the timing of revenues and expenses.

GAAP provides quite a bit of leeway on *how* processes are performed, as long as the key requirements are met. Consistency does not mean that we cannot change how we perform activities. Consistency does require a restatement of previous financial reporting periods when inconsistencies in results occur so that accurate comparisons can be made. In practice, most changes in process do *not* affect changes in accounting outputs or the results that are reported. It is the ease of comparing outputs or results of the process from period to period to which the Consistency tenet really applies. In the situation that the outputs are affected, a clear restatement can and must be made.

Interestingly, historically complex approaches to accounting processes often stem from the lack of timeliness of transactions; or in other words, the lack of flow. Companies design these processes to meet the requirements of Matching. When flow is significantly improved and lead times are dramatically reduced, you can consider other more efficient approaches.

Consider the typical manufacturer, with several months of inventory within its system at any time. What if you can significantly improve material flow through applying Lean concepts so that there is less than 1 month's worth of inventory in the system at all times? The typical accounting reporting cycle is 1 month. The expense associated *with* inventory will occur in the same period as the revenue generated *from* the use of the inventory (to make product, which is then sold to the customer). In other words, the expense and the revenue will always be matched as a natural result, with no additional time and effort required on the part of Accounting personnel. What if pull systems for materials have been implemented? The level of inventory for these items will fluctuate between two known values. Therefore, there is no longer the need to physically count on-hand inventory on a monthly or quarterly basis to verify quantities, as many companies presently do. Surely, you can reconsider your traditional approaches to inventory tracking, control, and valuation, and define new standard work that greatly simplifies the process.

In any organization, there are quite a few accounting processes. You need to focus on those that are currently consuming the most amount of people's time. To that end, you may need to first determine how people are currently spending their time. Collecting data for a period of time can help to focus the improvement efforts within the department. There are various approaches that you can take to reduce non-value-adding activities within each process. The details will depend on the circumstances, including the current systems you are using.

First, you need to map each process, most often using the value stream mapping tool. The value stream mapping tool includes key process metrics such as lead time, process time, some quality-related measures, etc. The current state map will identify numerous opportunities for improvement, though people do not always recognize them since they are so accustomed to the existing process. You need to challenge existing practices to streamline the process and create a future state map. Each accounting transaction must be challenged. The question you need to repeat over and over during this process is, "What information do we really need, and what is the easiest way (for everyone) to obtain it?"

For example, most companies have some form of "three-way match process." During a typical three-way match process, you will compare the information on a purchase order with information on the receiving document that arrived with the purchased materials, and again with information on the invoice received from the vendor. But for this process to work, you need to wait for all three documents to become available so that you can confirm the comparison. This can take several weeks or even months. If the information on any of these three documents does not exactly correspond or match the others, the vendor payment process is stopped, triggering an investigation involving accounting, purchasing, receiving, and inventory management personnel, as well as the vendor. But, often, the discrepancy between documents is insignificant, when you consider the overall dollar value of the item or items.

This process is meant to verify that the vendor is delivering the correct item in the correct quantity and at the correct price. Perhaps there is an easier way to determine this. Perhaps there is sufficient trust between the vendor and the customer so that each and *every* delivery does not have to be processed in this manner. Periodic audits of delivered items from each vendor may be sufficient. Perhaps a two-way match between invoice and receipt documents will be sufficient.

There are different ways to handle the transactions associated with the purchase and receipt of goods and services. What will work for a particular company will depend on several factors, the most important of which are the relationship between vendor and customer, the value of the items involved, and the level of visual management that has been implemented in purchased material storage areas. For example, it is feasible to identify incorrect shipping quantities by use of visual management techniques that clearly display standard order amounts and stocking levels. People responsible for inventory control can identify potential issues when they arise.

Table 7.1 Common Accounting Processes and Suggested Approaches

Process	*Suggested Approaches*
Labor reporting	Eliminate detailed labor reporting by operation Reduce the number of categories by which time can be reported Use electronic methods to capture labor-related data
Inventory valuation	For items on pull systems, use average inventory levels for valuation purposes. Actual inventory can be determined when required by law (annually for privately held companies, quarterly for publicly traded companies) Eliminate physically tracking work-in-process inventory once it has been significantly reduced
Budgeting	Reduce the overall lead time to complete the budgeting process from months to 1 or 2 weeks Reduce the number of accounts Eliminate detailed variance reporting
Financial reporting	Identify alternate means of generating required information that is not dependent on the month-end-close process Reduce the number of entities for which reports are generated Reduce the number of accounts Reduce the number of closes each year to what is required by law (annually for privately held companies, quarterly for publicly traded companies)

This is just one example, but there are plenty of others. A partial list of the most common processes is provided in Table 7.1, along with several suggestions to streamline them. Fortunately, there are now several excellent references on the subject of Lean accounting (*Practical Lean Accounting* Maskell, Baggaley, Productivity Press, New York, NY, 2004) that can provide greater detail on alternate approaches.

Making the Accounting Function Visual

Once again, all of our examples from Chapter 5 can be applied to the accounting function. As with all other functions, the ability to verify that the department is keeping up with demand is important. Accounting's visual management system must include a way to monitor various queues of work. Since the timing of activities is so important to this function, the

ACCOUNTING DEPT VISUAL MANAGEMENT

PLAN FOR EVERY PROCESS

ACTIVITY	RESPONSIBILITY	WHEN	STATUS
1. Process invoices	Anne	Daily by 3PM	●
2. Follow up on Aging Report	Jill	Tue/Thu by noon	○ refer to CI #2
3. Complete Inventory Analysis	Mike	Daily by 10AM	●
4. Post Cash	Anne	Daily by 1PM	●

KEY MEASURES WEEK OF: 4/5/10

METRIC	GOAL	ACTUAL	COMMENTS
1. DSO	35 Days	40 Days	refer to Aging Report
2. Invoice Accuracy	98%	99%	
3. Month End Close Lead Time	3 Days	4 Days	refer to CI Project #1
4. Audit Results	99%	99.5%	

CONTINUOUS IMPROVEMENT PROJECTS

WHAT	WHO	WHEN	COMMENTS
1. Warranty Allowance	John	By May Close	on Schedule
2. Cross Training on Aging Report	Jill	5/7/10	Anne: Complete Mike: open

MONTHLY ACTIVITIES MONTH OF ____

							STATUS
1	2	3	4	5	6	7	●
8	9	10	11	12	13	14	
15	16	17	18	19	20	21	
22	23	24	25	26	27	28	
29	30	31					

● = OK ○ = Non-Standard Condition

Figure 7.3 Visual management board in accounting.

accounting calendar, and the evidence that each activity has been performed as planned, needs to be made visible. See Figure 7.3 for an example of an accounting visual management board. Information on such boards should drive the continuous improvement efforts of Accounting personnel.

Improving the Accounting Function

Again, once time and capacity has been freed up through the use of Lean concepts, accountants can participate more fully in improvement efforts throughout the organization, bringing with them their analytical skills and systems knowledge. Accounting's role will change from simply providing reports on numbers to the organization, to driving improvement in the numbers themselves. An early focus might be efforts to improve the cash cycle, which consists of three elements: inventory, accounts receivable, and accounts payable. Although most Accounting departments are already involved in the collections and payables processes, Accounting personnel can also use their analytical skills to identify opportunities to reduce inventory levels.

These same analytical skills can be used to ensure that the company makes better decisions in investing in capital equipment, determining full acquisition cost from outside vendors so as to identify the best sources, in-sourcing or out-sourcing decisions, establishing target costs for new products under development, improving profitability of existing products, and assessing potential alternative service or manufacturing process changes.

Another way to improve efficiency in the accounting function is by leveling the workload within the department throughout each month, and even each year. There are plenty of opportunities within traditional month-end closing processes to level the workload. Again, approximately 50% of month-end related activities do *not* have to wait until the end of the month to be performed. They can be completed throughout the month in ways that can level the workload on the people responsible for performing them.

Customer Service

Most organizations have some type of Customer Service function, which typically performs several key processes. These processes usually include order processing or order entry and problem solving. Customer Service may also be expected to perform some sales activity and other secondary processes. In order processing, an order is received by some means (e.g., phone, fax, e-mail), and entered into a sales order system. Customer service associates also typically respond to customer inquiries of various types (e.g., order status, shipping information, pricing) and, in some companies, provide some technical support. It is a critical function, particularly in that it is often the "face" of the organization in the eyes of the customer.

Stability Issues with Customer Service

If customers are not satisfied with the service that they receive, then I can guarantee you that there is instability within the Customer Service function. Let us say that telephone response time is unacceptable or that orders are of poor quality—in other words, customers do not consistently receive what they ordered. Or maybe customers do not receive timely and accurate information when it is requested. In this type of situation, you need to address the stability issues first as part of your Lean customer service effort.

How you do this will depend on the nature of the issues. In the case of poor telephone response time, you might need to change the scheduled hours of customer service associates in line with actual call volume or even provide additional staffing. You might need to change the telephone "loops," or groups of people assigned to answer specific numbers. In some cases, you might need to update or alter the entire telephone system.

Further, you will need to identify and address the root causes behind order quality issues. These most often lie in the existing sales order systems. At one chemical company, orders were received electronically via an electronic data interchange (EDI) system. When you use these methods, you need to maintain a bridge between the customer and supplier systems over time. Unfortunately, in this instance, the system was not properly maintained and, over time, the quality of the incoming orders dropped to 12%. In other words, just 12% of all orders received by this method were complete and accurate. Data was lost during the transmission process or populated incorrect fields in the supplier system.

The EDI system relied on customer service associates to review each order, and to identify possible errors and correct them on a daily basis. However, this is the exact *opposite* of what should happen with these systems, since eliminating manual intervention is the intended benefit of EDI systems. In addition, the associates found it impossible to catch all of the mistakes. What this meant was that even after a great deal of manual effort, order quality still was just 90%. Let us look at this in real terms: the customer service department at this multibillion dollar company was staffed with 150 associates. If *all* EDI orders were properly received, approximately *half* that number would be required to meet demand. Interestingly, management was so focused on improving the 90% figure, it completely overlooked the non-value-adding effort necessary to reach this point. The technical problems with the EDI system had to be addressed early in the company's Lean initiative in Customer Service.

If you are using out-of-date online ordering systems or paper order forms, you need to correct and update them. In yet another example, customer service associates may not have access to the information that they need to rapidly respond to customer inquiries. Sometimes this is a relatively simple fix and only requires giving them access to particular systems or screens within systems. Other times, this can be a difficult undertaking.

At one company, the pricing policies were so complicated that it was nearly impossible to ensure that the correct price was entered when the order was processed. No sales order system in existence could have coped with the complexity of this company's pricing policies. The company employed a total of 12 "pricing specialists" to review orders and attempt to identify pricing discrepancies. Their process mainly involved checking with regional sales managers to confirm that the pricing was acceptable. But there were really no pricing policies at this company—*everything* was an exception. The company had to simplify their pricing policies as part of their Lean effort in customer service.

Whatever the nature of the instability, it needs to be addressed early in any Lean customer service effort. The good news is that when you effectively address stability issues, your customers will see real benefits early in the Lean effort. This can provide tremendous benefits to the organization, as customer satisfaction quickly increases. If the current level of customer satisfaction is acceptable, then the Lean effort can begin with standardizing existing processes.

Standardizing Customer Service Processes

Remember, the two main processes performed by Customer Service are order processing and problem solving, but there are also a number of secondary processes that customer service associates can perform in the course of any given day, week, month, or year. These include activities such as preparing literature for mailing to customers or outside sales resources, following up on payment issues, processing returned materials from customers, and generating reports. As I suggested for the previous functions, you might want to collect data for some period of time to identify the activities on which customer service associates are spending most of their time and focus your standardization efforts on these.

Regardless of how you receive orders, you need to streamline order processing. Detailed process maps can be created for each method in which orders are received (e.g., telephone, fax, postal mail, e-mail). Opportunities to simplify each process can be identified, and best practices agreed upon and documented.

Standardizing the problem-solving activity can be a bit more complex. First, you need to clearly identify the information required. Then, you will need to determine the source of the information (e.g., specific databases, paper files). Next, you need to develop ways to easily access the information. Perhaps the information can be made available to the customers themselves in some way, completely eliminating the need to involve a customer service associate. You need to work out a process for each and *every* type of information inquiry received from customers. Requests for shipping information might be handled in one way, while requests for historical order information in another. You may need to develop multiple methods for handling different types of customers—for example, Internet-savvy customers versus technology-resistant customers. Perhaps, with some education, customers who initially resisted particular methods will accept them. Customer satisfaction is paramount. Once again, the data you have collected on the nature and frequency of inquiries will help you to prioritize your efforts.

In terms of technical support, you need to develop tools and provide them to the customer service associates who are expected to provide the support. You can develop decision trees that reflect the knowledge of more experienced personnel. Decision trees define a series of questions that should be asked, and the responses to which lead to particular decisions or additional questions. These can be referred to by customer service associates as they work with customers to solve the most common technical problems over the phone. Similar tools can be provided to allow customer service associates to provide timely and accurate quoting information to customers.

Again, once you have standardized work, you will free up time of customer service associates that they can use to continuously improve service to the customer. But first, let us talk about how to make the function more visible.

Making Customer Service Visual

Much of the work that a customer service function performs is in electronic form. True, many companies still use fax machines, but it has been becoming less and less common over the years. The challenge lies in making the appropriate information visible. An example of a "plan for every process" for Customer Service was provided in Chapter 2 (Figure 2.2). Remember that "plan for every process" defined standard work for a customer service associate. It would be simple to add some visual indication to show whether or not each activity was completed as planned, providing the associates, supervisors, and managers with a simple means of frequently monitoring whether demand was met.

Some companies have made significant investments in information technology (IT) solutions to provide much-needed visibility in Customer Service. Sometimes the investment is warranted, but at other times simpler methods could have—and should have—been used. One challenge in using IT solutions is that IT approaches tend to require a common format for information and, in practice, this is not always possible. Do not forget that customers often have different levels of technological capability and know-how. Electronic Data Interchange (EDI) is a great example of the challenges associated with IT solutions. First introduced in the 1960s, there are *still* problems with this method of information exchange. The first attempt at

EDI standardization was made by the U.S. Transportation Data Coordinating Committee (TDCC) in 1968. In the 1970s, different industries created their own standards. This created problems since many companies operate in multiple industries. Attempts to create interindustry standards began in the 1980s with the American National Standards Institute (ANSI) X.12 protocol, which became the dominant standard over the past two decades. Although the use of EDI continues to expand beyond large corporations, it has been 40 years in the making and counting. Technology is not always the complete and timely answer.

Orders received by fax and e-mail can be made visible in different ways, and that is perfectly acceptable. The key is to develop a simple means of providing visibility. At the aforementioned multibillion dollar chemical company, customer service associates were asked to enter various process data into the existing system, capturing the frequency and nature of order quality problems. The data were summarized monthly and shared with other members of the department. They universally agreed that the information was not timely enough and that it took too much time for the associates to enter the data. For example, associates had to exit their current screens and go to another screen to enter the information. After trying a number of system-based methods, they came up with the simple solution of collecting the data manually on forms specifically designed for this purpose. Little writing was required, and there was no need for associates to change computer screens. Even for a multibillion dollar business with 150 customer service associates, manual methods worked better.

Fortunately, telephone systems have had great capabilities in this area for years. A wide variety of software packages can monitor telephone response time, the percentage of dropped calls, and other useful statistics. Further, some of this information can be tracked in near-real time, which makes it more meaningful. Again, your challenge is to make this information visible so that it drives appropriate decision making. Many companies display this information on monitors, sometimes as part of a pull system such as the one described in Chapter 4.

Once adequate visibility is in place, we can focus on continuous improvement in the customer service function.

Improving the Customer Service Function

Well-implemented visual management techniques can help you quickly and effectively identify opportunities for improvement in the customer service function over time. These techniques can help you improve call response time and enhance associates' problem-solving abilities through additional as needed. Better still, customer service associates can participate in problem *elimination* efforts. Too often, customer service personnel are not permitted to participate in kaizen or other improvement efforts, because of a perception that they cannot be away from the phones for any significant amount of time, if at all. However, customer service associates provide an essential perspective for many improvement efforts. The very nature of their role provides ready access to the voice of the customer. Again, the streamlining benefits realized during the development of standard work will free up time for them to participate in Lean improvement efforts both within and outside the customer service department.

At an automotive parts manufacturer and distributor, a customer service representative participated in a twice-weekly review of a visual problem board located in the accounting department. On this board, accounts receivable personnel identified issues with invoicing, collections, etc., by customer and as they were identified. The review took about 15 minutes to complete, and responsibility for attending was periodically rotated among the customer service associates. The customer service representative at the review then returned to the department, contacted the appropriate person, and formulated a response to the issue before the next review. Customer Service often had important information regarding the issue. This simple collaboration between departments allowed for the timely resolution of issues and helped reduce the days sales outstanding (DSO) from 73 to 70 days, significantly impacting working capital in this $1.5B per year business. Previously, these issues went unaddressed for 30 or more days. Customers were also pleased that these issues were resolved in a timely manner with minimal effort on their part.

Another common improvement effort in which customer service associates can participate is proactive selling. Remember, customer service personnel are sometimes involved in sales activities but, unfortunately, they do not always have much time for it, as they struggle to keep up with the demand for order processing—a reactive form of sales, as the sale has already been made. But once we streamline order processing and other secondary activities, customer service associates will have more time available for proactive selling. For example, they can contact existing and potential new customers to generate sales that otherwise would have been missed. In several companies implementing Lean in customer service, 45 to 60 minutes each day were made available for proactive selling, and sales increases of as much as 6% were attributed to this new capability.

A common obstacle to a Lean effort in Customer Service is the existing measurement systems. In fact, in some organizations, existing measurement systems are the primary cause of great instability in the customer service function. For example, a certain technology company was widely known for its poor customer service, with lengthy call wait times. Customers were passed from representative to representative, and problems went unresolved after *numerous* hours on the telephone. The company's reputation suffered horribly. What were the root causes of the problem? Customer service personnel were measured on call duration (not including queue time), and the shorter the time spent on the telephone with the customer, the better. Associates were also measured on call volume—the greater the number of calls taken, the better. Not a single measure had any relationship to actually solving the customer's problem. Customer service associates learned to game the system, repeatedly placing customers on hold to reduce the call duration time. Transferred calls counted in the call volume metric, so associates were able to both get off the call quickly and increase their call volume. All of these undesirable behaviors—and then some—were actually created by the measurement system.

This company needed to change the measurement system at the beginning of the Lean effort in order to affect the proper behaviors and achieve some amount of stability. When making changes to a measurement system, tread carefully. In the case of Customer Service, the wrong measurement system can lead to near disastrous results, since Customer Service is the public face of the organization. If your goal is continuous improvement, then you need to establish measures that drive appropriate improvement efforts for each function and for the organization overall. We will discuss measurements in more detail in Chapter 8.

Human Resources

The real purpose of human resource (HR) departments is to continually develop the capabilities of an organization's human assets so that it can meet its objectives over time. Unfortunately, HR departments in most companies get caught up in regulatory and compliance activities, such as making sure that people are paid accurately (payroll), reporting required information to local, state, and federal agencies, and administering benefits programs. This leaves little time for true organizational development activities. Your Lean effort in this area needs to focus on streamlining transactional processes so that time can be freed up for other activities through which the HR professional can add greater value to the organization and its members. This is equally true for an HR department of one and for entire departments with multiple people.

Stability Issues Relating to Human Resources

In most organizations, many HR transactional processes are well defined and consistent. These include personnel record keeping, disciplinary practices, performance evaluations, and the like. Interestingly, though this is often *not* the case with orientation and training processes for new employees; it is often left to other functions to determine the best way to perform these important activities. The result is a great amount of variability, leading to lack of standardization, lost productivity, poor quality, and even high turnover.

If in fact turnover or retention is a major concern, the early focus of your Lean HR effort must be on this disruptive and costly issue. The cost of replacing a single individual can range from approximately $4,000 to $100,000+ or from 25% to 150%+ of the individual's annual salary. Elements of the cost of turnover include separation or exit costs, hiring costs, orientation and training costs, productivity loss, and lost expertise. The specific cost of turnover will depend on the nature of the position and the industry. According to a U.S. Department of Labor, Bureau of Labor Statistics (BLS) study, the national average for turnover for the year ending August 2006 was 23.4% overall, and 26.5% for private industry.

Regardless of the actual figure, turnover is costly to any organization. Further, it can significantly hamper Lean efforts, since HR personnel will spend most of their time simply "filling a slot," rather than on true organizational development activities. Break the cycle! At one company of 400 employees, the turnover rate was 50% for production associates. The estimated cost to replace each production associate was $5,000, putting the annual cost of turnover at approximately $1,000,000. In exit interviews, 40% of departing associates cited a lack of training as a primary reason for leaving. Other concerns included safety (also related to a lack of training) and payroll and benefit issues (related to a poor orientation process). The organization established a training center staffed by two full-time employees and equipped with older but still suitable production equipment.

The organization developed a 40-hour curriculum for the training center, including orientation to the company, its policies, products, and customers. Newly hired associates were trained to operate equipment in a controlled environment, allowing the trainers to ascertain the particular skills and abilities of individuals. This knowledge allowed individuals to be placed in the positions most suitable to them. The final 4 hours of the curriculum included an introduction to the area that associates would be assigned to, and they were given time to actually observe the work environment. The results were impressive, with turnover reduced by 50%. The investment in training was estimated to be approximately $100,000 per year (one-week of training for each new employee, plus the cost of two full-time trainers). The savings were estimated to be $500,000, for a net savings of $400,000 per year.

The lack of standard orientation and training is inexcusable. But, by now, you will know what the standard excuse is: not enough time. But again, much more time is lost due to a lack of standard orientation and training. Remember our discussion of the concept of job instruction (JI)? This concept must be the foundation for all orientation and training programs. The specific content will vary based on the role or position, but the general approach will be the same. Defining training content, ensuring that skilled training resources are available when necessary, and providing an effective learning environment are all key here.

You need to identify the root causes of turnover. In our example, the principal cause was poor orientation and training processes. However, there can be other causes, and each needs to be addressed. Often, these are related to HR processes that involve supervisors and managers, such as performance evaluations, and the administration of disciplinary policies. Developing and standardizing these important processes may provide much-needed stability.

Standardizing HR Processes

Our goal is to streamline the overall "hire-to-retire" value stream and all activities associated with it. For example, we should not spend too much time on payroll processes; given the available technology, payroll processing should be practically automated. The same can be said for local, state, and federal reporting of labor information. If members of the organization are spending substantial time on these processes, that is where early Lean HR efforts need to be focused.

Performance evaluation is a very important HR process—people want and need to know how they are doing, and performance needs to be documented periodically. If properly conducted, the process provides an excellent opportunity to improve the capabilities of individuals and, in turn, of the organization. Unfortunately, performance evaluation processes in some companies have become overly complex. At one company, the overall lead time for the personnel performance evaluation process was 6 months. The company's approach was well intended and was meant to be fairly exhaustive, soliciting input regarding an individual's performance from as many as two dozen fellow employees. However, the administrative burden this placed on HR personnel was intolerable. The organization simplified its process by setting reasonable guidelines and limiting the number of people solicited to provide input. Evaluation documentation was cut from as many as 12 pages to 2. Once the new process was in place, it was generally agreed that there was no degradation in the quality of the feedback, and the lead time was reduced to 1 month.

In other companies, performance evaluations are not consistently performed, either in timing, format, or the quality of the feedback. Nonstandard forms may be used across the organization, and rating systems, if they are used at all, are not uniformly applied. Many managers put off completing evaluations, in part because of the "batching" nature of the evaluation process. Many companies time the process to coincide with the end of a calendar or fiscal year to link compensation actions with an annual budgeting process. As we discussed in previous chapters, you need to challenge *any* form of batching. The quality of evaluation feedback deteriorates when large batches are processed, because supervisors and managers are rushed to complete the large batch by the designated deadline. Many companies address this by performing evaluations on or around the employee's anniversary date of hire. This usually results in some amount of "leveling" of the process throughout the year. Once you have simplified the process and associated documentation, you need to train all members of the organization in the new standard process.

As I suggested during the accounting discussion, you might want to have HR personnel and other members of the organization track the amount of time that they spend on HR processes. This information can help you prioritize your improvement efforts to streamline and standardize.

Making the HR Function Visible

Again, you need to set up a visual board to display all HR activities, such as processing payroll, generating reports, processing performance evaluations,

HR DEPARTMENT VISUAL MANAGEMENT

PLAN FOR EVERY PROCESS

ACTIVITY	RESPONSIBILITY	WHEN	STATUS
1. Process Payroll	Sue	Thurs by 2	●
2. Change in Benefits Requests	Bob	Wed by 4	●
3. Wage & Salary Changes	Sue	Tue by 1	●
4. Safety Meeting	Jane	Every 3rd Thurs.	●

KEY MEASURES WEEK OF: 4/12/10

METRIC	GOAL	ACTUAL	COMMENTS
1. On-Board LT	10 Days	10 Days ●	
2. On Time Performance Eval	100%	100% ●	
3. Employee Satisfaction	4.5	4.3 ○	Refer to Dec 09 Survey
4. Cross-Training "Barometer"	85%	80% ○	Refer to CI Project #1

DAY OF: 4/14/10 **TEAM STATUS**

NAME	IN/OUT	SKILLS MATRIX		
		PAYROLL	BENEFITS	ON-BOARD
Sue	IN	X	X	X
Bob	IN (OUT FRI)	X		X
jane	IN	X		X

CONTINUOUS IMPROVEMENT PROJECTS

WHAT	WHO	WHEN	COMMENTS
1. Develop Lead plan	Bob	Aug '10	on Schedule
2. Cross Train on Benefits	Sue	Jul. 15	on Schedule
3. New Benefits Program	Sue	Jul. 30	on Schedule

● = OK ○ = Nonstandard Condition

Figure 7.4 Visual Management Display in HR.

conducting exit interviews, etc. And again, the board should include a visible indication when each activity has been completed on time. A section of the board should list all identified continuous improvement or organizational development projects and the status of each. An example is provided in Figure 7.4.

Improving the HR Function

Remember, the HR department in a Lean enterprise proactively identifies organizational needs (e.g., new skills, additional technical know-how) and continually improves organizational capabilities. You need to periodically assess your current organizational capabilities and compare those to what is presently needed or will be needed in the future. You need to establish and maintain some form of skill matrix that identify required skills and the present level of capability for each. It does not have to be centrally and electronically stored; instead, HR professionals can visit different work areas and determine training needs by reviewing visibly posted skills matrices (see Figure 7.5).

HR needs to have mechanisms in place to capture the voice of the customer, for example, periodic employee surveys. Today, surveys are very easy to design, deliver, and analyze using cost-effective online services. Exit interviews are another means of capturing important information. But remember

Name	Registration	Insurance	Discharge
Bob	X	X	
Sue	X		X
Jane	X		
Eileen	X	X	X

Figure 7.5 Skills matrix example.

that these surveying techniques are very reactive, and HR needs to *proactively* identify the knowledge and skills required to support the business in the longer term. To this end, HR management must actively participate in the company's strategic planning processes; this will keep them abreast of the company's direction. In turn, HR personnel will be able to identify gaps in organizational capabilities and can identify outside resources to fill them.

HR professionals can participate in, and even facilitate, kaizen events. HR professionals often have important interpersonal skills that can be invaluable during such events. Eighty percent of the success of any change can be attributed to behavioral factors and 20% to technical factors. Technical factors are the specific improvements to be made or solutions to be implemented. Behavioral factors include the buy-in and commitment on the part of the people involved to make the changes work. Who better than HR professionals to address behavioral issues that come up during improvement efforts?

Other business improvement efforts include visiting other companies to "benchmark"—in other words, to gauge business practices within other organizations—and, more generally, to learn. HR personnel can use their affiliations with professional societies to identify these opportunities. In fact, maintaining an external perspective is important to prevent an organization from becoming too insular.

One of the most important functions of the HR department is the ongoing development of the organization's leaders. Remember, the goal of a Lean enterprise is to create a culture of continuous improvement. Since leaders define culture, HR professionals must ensure that they have the requisite skills to successfully fulfill their role. Too many companies simply hire or promote a person into a leadership role but do not prepare that person for the extent of his or her responsibilities. It is risky to assume that the person has the necessary skills, or will somehow develop them over time. Further, individuals currently in leadership positions do not always possess the needed skills to be successful in the long term. You need to address both of these situations, either with HR personnel directly providing the necessary education to new or existing leaders or arranging for other means to do so. This often represents a very different role for a traditional HR professional and a different relationship between HR and leaders within the organization. You need to treat the HR function as an equal partner in leading the business, because they can be real change agents in an organization.

Summary

It is impossible to cover *all* functions across all organizations, but this general approach—Stabilize, Standardize, Visualize, and Improve—can be applied to *any* function. Remember, your starting point will depend on your current conditions. They will dictate whether you need to start with Stabilize or Standardize or, on some occasions, Visualize. You will need to collect data to help prioritize your standardization effort and use the questions in Chapter 5 to develop an appropriate visual management system.

Your real difficulty will likely lie with Improve, since it may require significant changes in people's roles, and those are not always easy for people to make. In most cases, you will need to provide lots of leadership support, and possibly education and training. Often, these changes in roles require new skills or, at the least, a refresher course for skills that have long gone unused. Supportive leadership is critical. In the next chapter, I will show you how to lead the Lean office or service organization.

Chapter 8

Leading the Lean Organization

By applying the concepts covered in Chapters 1 through 6, you will implement highly performing, controlled processes that are, to a large degree, worker managed. When processes are worker managed, the role of the *manager* in a Lean organization changes, from manager to *leader*. When you manage according to *Merriam-Webster*, you "handle" or "make and keep compliant." But when you *lead*, you "guide on a way, especially by going in advance." In other words, leaders get people to move to a new or different place, often when they did not want to believe, or did not believe, it was possible. When you implement Lean in your organization, your traditional management role will change from one based on control and direction of basic tasks to leading your people and organizations to a new place—a continuously improving organization, with a more enriching and rewarding work environment.

You might think that this is unrealistic. Or maybe you or your staff do not want this kind of work environment. Maybe you think that people only care about being paid. But let us take a look at Abraham Maslow's seminal and widely accepted hierarchy of needs, from his paper "A Theory of Human Motivation" (1943) (see Table 8.1). Lower-level needs, such as physiology and safety, must be fulfilled before higher-level needs can be met. In other words, if you feel unsafe in your work environment, you are not likely to want anything to do with some problem-solving effort. However, once your lower-level needs are fulfilled, you are more likely to accept, and even embrace, an opportunity to fulfill higher-level needs, such as "esteem" and "self-actualization." So, what does that mean for your Lean organization? Well, aspects of higher-level needs include achievement, problem solving,

Table 8.1 Maslow's Hierarchy of Needs

Self-actualization	Creativity, problem solving, lack of prejudice, acceptance of facts
Esteem	Confidence, achievement, respect of and from others
Love/belonging	Friendship, family, etc.
Safety	Security of body, health, employment, resources, property, etc.
Physiological	Water, food, sleep, breathing, etc.

confidence, and creativity—and *that is* what you need to develop a culture of continuous improvement. These also include respect, Lack of prejudice (which strongly relates to respect), and acceptance of facts. An organization's members also require these needs to be met—and who do you think can do that? The organization's leaders. People *need* what the Lean business model offers.

Still not convinced? Let us take a look at Frederick Herzberg's book *The Motivation to Work* (Wiley, New York, NY, 1959), which identified both work characteristics that lead to satisfaction and those that lead to dissatisfaction (see Table 8.2). In Table 8.2, the left column identifies characteristics that lead to dissatisfaction with the work environment; these are presented in order of importance, from higher to lower. If you want to upset people, and fast, institute a company policy that they do not agree with. Note that "supervision" and "relationship with boss" are the second- and third- highest factors that lead to work dissatisfaction. Also note the position of "salary": fifth. These factors, at a *minimum,* need to be attended to so that members of the organization do not become dissatisfied. It is important to realize that these factors *do not* create satisfaction. When attended to, they simply bring people to a "neutral" state that is necessary to allow them to become satisfied by other factors, as identified in the right-hand column in Table 8.2. For example, consider salary. Although money cannot buy happiness, it can provide a sense of security.

The right column of Table 8.2 lists the characteristics that lead to satisfaction with the work environment, again presented in order of importance from higher to lower. Note that "achievement" and "recognition" are at the top of the list. You might be surprised by this list, but if you consider Maslow's levels of needs, particularly "esteem" and "self-actualization," it should begin to make sense. You should also note the absence of "salary" from the list. But why would not salary lead to satisfaction, when it can lead to *dissatisfaction*? That is because an acceptable salary—one that fulfills a staff member's physiological and security needs—simply brings them to the point of *not* being displeased.

Table 8.2 Herzberg's Job Enrichment Characteristics

Leading to Dissatisfaction	*Leading to Satisfaction*
Company policy	Achievement
Supervision	Recognition
Relationship with boss	Work itself
Work conditions	Responsibility
Salary	Advancement
Relationship with peers	Growth

Note: Presented in order of higher to lower importance.

It is damning with faint praise, and more is needed to make the work environment satisfactory, enriching, and rewarding, as noted on the right-hand column of Table 8.2. Since Herzberg's initial publication of the list in 1959, these factors have become widely accepted and supported by numerous studies. Although the order of importance in each list can change a bit from study to study, the characteristics themselves have remained constant. These factors are completely aligned to the Lean philosophy, which means that employees are likely to find the culture in a Lean organization to be more satisfying.

If Maslow and Herzberg's work makes sense to you, then the basis for your new role becomes clear—to drive continuous improvement. In this chapter, we will talk about how you can do that by examining the following topics:

- Driving continuous improvement (PDCA)
- Mentoring
- Going to the gemba
- Performance measurement
- Recognition

Together, these topics define the knowledge, skills, and methodologies that leaders need to be successful in the Lean organization.

Driving Continuous Improvement (PDCA)

First things first: every leader and every organization needs a methodology for continuous improvement. Fortunately, a methodology exists that has stood the test of time for more than 50 years: W. Edwards Deming's

"Plan-Do-Check-Act" (PDCA) improvement cycle. Deming introduced this concept in the 1950s, but its origins date back to Dr. Walter Shewhart in the 1920s. And *all* subsequent improvement methodologies can be traced to "PDCA." To lead effectively, you need to understand the PDCA cycle and consistently put it into practice. Let us provide a quick review of the methodology.

In the "plan" part of the cycle, you will identify a problem (or problems) and develop an understanding of its impact on a process and the business as a whole. To do this, you need to collect relevant data and analyze the data to develop a clear statement of the problem. As Charles Kettering, an American inventor, once said, "A problem well stated is a problem half solved." Once a problem is well understood, you need to identify the "root causes," or true origins, of the problem. In most cases, there are multiple process-related causes that give rise to a problem. As a leader, you need to practice the "5 Whys" (see sidebar) and challenge the assumptions related to root causes. By asking "why," your initial gut feeling of possible causes gives way to more thoughtful consideration. Once you have identified possible root causes, you can identify solutions or countermeasures that address them.

SIDEBAR The 5 Whys

The best question that a leader (or anyone, for that matter) can ask is "Why?" You might be reluctant to do so (for fear of appearing less smart, or concerns over challenging others). But reluctance only stands in your way, and you need to overcome it, because asking why is welcomed and encouraged in the Lean organization. It is nothing but an attempt to obtain more knowledge, which is exactly what you want to happen. Repeatedly asking "why" is referred to as the "5 Whys" in Lean organizations. Using the 5 Whys, you do not simply accept the first response but attempt to identify more thoughtful explanations. Let us look at a classic example, involving the breakdown of a piece of equipment. "Why did the machine break down?" The first response might be something like "Because there was an overload and a fuse blew on the motor." If you stop there, the solution or countermeasure would simply be to replace the fuse. But what if we asked a second question instead? "Why did the fuse blow?" The investigation prompted by this question reveals that the motor bearing was not lubricated properly. This, in turn, will prompt another question, "Why wasn't it lubricated?" The response this time is that "the lubrication pump was not operating properly." Our next question is, "Why was the pump not operating properly?" By this point, our investigation reveals that the pump shaft had worn. If we stop here, our solution would be to replace the shaft. However, one more question, "Why did the shaft wear?" identifies the true root cause: "Because a strainer was not properly replaced and metal chips made their way onto the shaft, wearing it down." Short of addressing the root cause involving the strainer, any solution implemented would represent a temporary fix, and the problem would reoccur. It does not always take 5 whys. Sometimes the root cause can be identified after fewer questions, and at other times more are required. This is a powerful, yet simple, practice, and it can make a big difference in your organization.

The result of the "Plan" step is typically a list of agreed-upon solutions or countermeasures, and you will need to put them to the test. This happens in the "Do" and "Check" steps. But before we move on, I need to emphasize that there is a lot to be completed in the Plan step. You, the Leader, need to be patient to ensure that this step is adequately completed. Too often, leaders want to skip to "Do," jumping to action based on little or no information, and with an insufficient understanding of the problem and current situation. This often happens under the mistaken belief that quick action is a characteristic of effective leadership.

During the "Do" step, you will implement your assumed solutions, by developing and monitoring an action plan for successful implementation. Your action plan will include the three "W's"—What, Who, and When. *What* specific actions need to be completed to implement the solutions? *Who* is responsible for seeing that they are successfully completed? *When* will each action be completed? As a leader, you need to make sure that all of these questions are answered when the action plan is developed. There has to be follow-through on *all* identified actions. Nothing less should be accepted. Leaders must assist associates in overcoming any obstacles that may arise.

The "Check" step comes next. You can never assume that the solutions or countermeasures implemented will be successful. Instead, you need to verify their effectiveness. Often, this means additional data collection. Once again, patience is needed as the impact of changes may take some time to assess. Too often, leaders and fellow team members are quick to move onto the next problem, foregoing this step altogether. You need to avoid this. Initially, Deming had wanted to call this step "Study," as it provides a better description of its purpose. Commonly, the solutions implemented in the "Do" step are really just tests of a hypothesis that are not fully implemented until proved effective. Full implementation will happen in the next step, "Act."

During the "Act" step, you will take actions to turn your changes into your new standard (again, once they have been proved effective). During this step, work instructions are updated, all affected people are trained, and control plans are put in place to ensure that the changes are sustained over time. Do not overlook this step. If you do, I can almost guarantee that your problem will be back.

Every leader within a Lean organization will personally practice PDCA whenever the opportunity arises. When a problem is encountered, the leader must follow the PDCA process in a disciplined manner. He or she also needs to encourage others to do the same, and lend assistance and guidance until that point in time that it becomes second nature for them. A detailed methodology for the PDCA improvement cycle, in the form of Job Methods, or JM, appears in the sidebar.

SIDEBAR Job Methods (JM)

You will remember that in Chapter 2, I introduced the term "Training Within Industries (TWI)." Again, TWI was developed in the United States to support the war effort during WWII. TWI comprises the three "J's": Job Instruction (JI), Job Relations (JR), and Job Methods (JM). With JM, you can improve a process by making the best use of people, equipment, materials, etc. The steps are

- Break down the job. List all of the details of the job *exactly* as they are currently done.
- Be sure that the details include material handling, machine work, handwork, general organization, information required to perform the work, and time to perform tasks.
- Question every detail. Why is it necessary? What is the purpose? Where should it be done? Who is best qualified to do it? What is the "best way" to do it?
- Question the use of materials, machines, equipment, tools, product design, layout, workplace organization, safety, and housekeeping.
- Develop your new method—eliminate unnecessary details, combine details when practical, rearrange for better sequence, simplify all necessary details.
- Make the work easier and safer, pre-position materials, tools, and equipment at the best places in the proper work area. Use gravity whenever possible; let both hands do useful work; use fixtures instead of hands for holding work.
- Identify who will be affected by the new method. Work out the ideas with others.
- Apply the new method. Sell the new method to others. Explain the advantages.
- Apply Job Instruction (JI) techniques for the new method to teach it to the initial workers expected to try the new method.
- Verify that the new method is working. Check quality, safety, and speed of the new method.
- Give sufficient time after training has been given for people to become accustomed to the new method.
- Share results with other workers to reinforce the new method.
- Use the new method until a better way is developed.
- Teach other workers the new method using JI techniques.
- Document the new method in a simple and visual way. Post in the immediate area.
- Make all physical changes necessary to support the new method—change layout, rearrange workstation, etc.
- Use visual techniques to clearly show the new work area arrangement.

Mentoring

Leaders in a Lean organization mentor others—they coach, counsel, guide, and tutor them. In other words, the Lean leader needs to be a teacher, taking advantage of *all* teaching opportunities. This role might be uncomfortable for a traditional manager, but the more knowledge and ability the team has, the easier it is on the leader. And this task can be as simple as sharing information with co-workers when opportunities arise; for example, by explaining business decisions that have been made, and how, or by sharing information about an aspect of the business that people do not normally have access to. Most teaching does *not* involve standing in front of a classroom of people. Again, Training Within Industries—this time using Job Instruction—can help you attend to this leadership aspect (see sidebar).

SIDEBAR Job Instruction (JI)

The second component of TWI is Job Instruction (JI). JI was developed to facilitate the transfer of skills to people in a production environment. However, the underlying concepts are applicable to teaching of any kind.

Here are some things to consider when preparing to instruct:

- Have a timetable that details the degree of skill you expect the person to have, and by what date.
- Break down the job. List important steps and identify *Key Points*. Key points are the "how and why" of the job. Safety is *always* a key point.
- Have everything ready: the right equipment, material, and supplies.
- Have the workplace properly arranged, just as the worker is expected to keep it.

In addition, there are steps to follow *while* instructing:

1. *Prepare the associate*. Put him or her at ease. Talk about the job and find out what he or she already knows about it. Get the person interested in learning the job. Place the person in the correct position to observe.
2. *Show the associate how to do it*. Tell, show, and illustrate, one important step at a time.
3. *Explain Key Points*. Identify the key quality aspects of the job; in other words, what constitutes acceptable and unacceptable quality. Explain *why* they are important. Identify the key safety aspects of the job, or *how* to perform the job in a safe manner. Instruct clearly, completely, and patiently, but do not provide more information than he or she can handle.
4. *Show the associate how to do it again*. Let the associate watch you do it again. Repeat Key Points as you perform the job.
5. *Let him or her perform simple parts of the job*. Ask the associate to explain each Key Point to you as he or she does the job again. Identify and correct any errors.
6. *Let the associate perform the whole job*. Have the associate explain each Key Point to you as he or she does the job again. Make sure that the associate understands. Continue until YOU know that HE or SHE knows.
7. *Put the associate on his or her own*. Designate a point person for help (yourself or a co-worker). Check frequently. Encourage questions. Taper off extra coaching and close follow-up.

These are the three cycles of learning. The first cycle is to observe; the second is to participate, at least partially; and the third is to perform the task completely with full understanding of the key points. Again, this approach, which has been proved effective for more than 60 years, can be applied to teaching of any kind. Leaders need to use some form of the approach, and use it patiently; you cannot take shortcuts for expediency's sake, because it will undermine the effectiveness of the instruction. Always remember, if the associate has not learned the task, the instructor has not taught it.

Job Instruction (JI) provides a framework for teaching. But what about the subject matter? What should the leader teach? Obviously, the logical first step is to teach the ways that required work should be performed within individual departments or functions. Once you have implemented standard work, you will find that instructing employees on operations-related activities can be done very effectively and efficiently. So, that is the obvious subject matter. What else should the leader teach? Next, you need to provide instruction through the actual application of the "Plan-Do-Check-Act" improvement cycle. Leaders should be developing the problem solving skills of others in their areas of responsibility by working with people to solve actual problems. Further, the leader needs to use proven basic quality management tools while doing this (see Appendix). In this way, people will gain an understanding of these tools and their use, and will then be able to apply them on their own in the future as opportunities arise. This will make them more effective problem solvers.

More generally, leaders need to help develop the critical thinking abilities of their staff. Of course, the leaders themselves must develop these abilities first. How? First, let us look at the definition of "critical thinking."

- In their book, *Critical Thinking: Its Definition and Assessment* (Springer Netherlands, 1997), Alec Fisher and Michael Scriven define critical thinking as "skilled, active, interpretation and evaluation of observations, communications, information, and argumentation."

In other words, critical thinking is the *purposeful* and *reflective* judgment about what to believe or what to do in response to *observations*, experience, verbal or written expressions, or arguments. It gives due consideration to the evidence as judgments are made, the relevant criteria for making the judgment well, the applicable methods or techniques for forming the judgment, and the applicable theoretical constructs for understanding the problem and the question at hand. Critical thinking employs not only logic, but broad intellectual criteria such as *clarity*, *credibility*, *accuracy*, *precision*, *relevance*, *depth*, *breadth*, *significance*, and *fairness*.

It is not difficult to understand the meaning of critical thinking (to clearly and objectively look at a problem or issue from different perspectives), nor its intent (to effectively address the problem or issue). In practice, however, it *can* be difficult. As I have already noted, however, leaders must be patient in order to practice critical thinking. Please do not confuse patience with inaction, but you cannot be too quick to take action without the necessary facts. How can you develop these skills, in yourself and in others?

You can only develop these skills through practical application within the context of the PDCA improvement cycle. Further, if you will remember the basic methodology of Job Instruction (JI), it will take approximately three cycles of application, no less, to effectively transfer the requisite knowledge and skills.

SIDEBAR The A3 Management Process

A subject gaining renewed interest is A3 reports, or storyboards. A3 refers to the size of the single page of paper (approximately 11″ × 17″) on which a story is conveyed. Creating a one-page storyboard with multiple sections that closely mirror the PDCA process was one of the quality management techniques espoused by W. Edwards Deming, among others, as early as the 1960s. Unfortunately, along the way, A3 storyboards became viewed as simply a communication technique to succinctly convey a story. The original intention—to ensure adherence to the PDCA improvement cycle—was lost. Managers could confirm that a process improvement team was effectively addressing a problem (i.e., practicing critical thinking) simply by reviewing the storyboard and could then provide guidance as necessary. This is referred to as the "A3 Management Process." The manager may suggest that the team gather more data to better understand a problem or involve particular individuals to obtain their perspective and suggestions for improving the effectiveness of the team in addressing the problem. This way, the leader can mentor an individual or group of individuals to improve their problem-solving abilities. John Shook addresses this important subject in his book, *Managing to Learn: Using the A3 Management Process to Solve Problems, Gain Agreement, Mentor and Lead* (Lean Enterprise Institute, 2008).

Mentoring often occurs during one-on-one interactions between the leader and an individual. Let us say that an individual needs to improve his or her skill in a specific operations activity. Instead of delegating this important instruction opportunity to others in the department—which is what a traditional manager is likely to do—the leader should work directly with the individual to improve that skill over time. Often, managers delegate these tasks because they believe that the people already performing the activity every day can do it better or faster than they can. But an effective leader needs to know the work, though he or she does not need to be as proficient as others in order to effectively teach it. When you delegate instruction to an associate's peers, for example, the effectiveness of the teaching can actually *decrease*. For example, the person expected to deliver the instruction may be occupied on other matters, and may see teaching as a disruption. Just as important, a critical mentoring opportunity will be lost. Again, leaders should be primarily focused on continuous improvement. What better way to ensure continuous improvement than by focusing on the one-on-one development of an employee?

Going to the Gemba

Leaders in the Lean organization should always look for opportunities to "go to the gemba." Gemba (also spelled genba) is a Japanese term for "actual place," often referring to the place where value-creating work is performed. Traditional managers have gotten away from this simple practice, instead spending their days in meetings rather than in the trenches. "Going to the gemba" is an easily implemented practice. Let us say that you are discussing a problem in a meeting or conference room. Stop the discussion and take the group "to the gemba," to the actual place where the problem occurs. This way, your group can directly observe the problem and the current conditions in the work environment that give rise to the problem. Most often, the solution to the problem can also be found at the gemba.

As a leader, you can periodically perform "gemba walks," in which you and others can generally observe the work environment. The focus of your walks can vary, from the general flow of some type of information or service, to identifying non-value-adding (NVA) activities or waste in the process, to general workplace organization. Do not try to focus on everything during a particular walk, even if you are an experienced gemba walker. Select a single focus prior to the walk.

Gemba walks provide an opportunity for the leader to act as a mentor. Consider having a person (or multiple persons) join you on the walks. For example, if you take a team on a gemba walk focused on NVA activities or wastes, it is a golden opportunity for you to review the eight categories of waste in real time, using real-life examples. This will help your team create "eyes for waste," a valuable ability in the Lean organization. Or you could take advantage of a walk focused on workplace organization to help participants develop a deeper understanding of 5S concepts.

During a gemba walk, leaders should periodically review the elements of the visual management system that has been implemented in the workplace (as described in Chapter 5 and elsewhere in the book). When associates see the leader taking an interest in the information posted as part of the visual management system and asking questions about it, it reinforces its importance and ensures that it will be sustained over time. An added benefit is that associates in any given department will no longer need to spend as much time periodically developing reports for review by management, as the managers—now leaders—will obtain the necessary information through the visual management techniques employed. Leaders will also benefit from this activity by being kept aware of things in more efficient ways, thereby needing to spend less time in department meetings. Going to the gemba, or what was described years ago as "managing by walking around," is a simple practice that can provide important benefits to leaders in Lean organizations.

Performance Measurement

No leadership discussion would be complete without a discussion of performance measurements. After all, the role of the leader in the Lean organization is to drive continuous improvement, but improvement in what performance measures? Measurements drive behavior—even dysfunctional behavior—in most work environments. You need to carefully consider which performance measurements you are going to use. The specific measures that you choose will depend on the particular department or function but, regardless of which specific measures you use, there are a few things that you need to keep in mind.

First, remember that fewer are better with respect to the number of measurements. There should be little reason to go beyond key measurements of quality, service, safety, and cost, which apply to almost every process. Second, consider the behaviors that you want to affect. You need to identify and develop measurements that drive the correct behaviors, such as a strong focus on meeting the needs of internal and/or external customers. You will also need to consider the key Lean principles of value, flow, pull, and perfection. Quality, Service, Safety, and Cost provide a good framework for discussion to help identify specific measures since they have broad application to most office and service processes.

You need to add some measure of service to the customer to your performance measures. Maybe you could track the percentage of "on-time" completion of key activities that directly affect the customer: for example. Did department personnel complete activities within the desired timeframes to ensure a high level of customer satisfaction with the services provided? This same metric can often serve as a measure of flow. Remember that you will need to balance all of these metrics with a quality measurement. It is important to meet the quality needs of the customer as well as timeline expectations. Together, these metrics can serve as a measure of the key objective of pull: did the customer receive what they wanted, when they wanted it? The point here is that by selecting a few key measures, all of the Lean principles can be covered.

Often, organizations will include a measure relating to the condition of the workplace (safety, general organization). This can be a numerical score summarizing the condition of the work environment and can be generated during a 5S audit. Since perfection is the goal, you might develop a measure to track the number of ideas for improvement that were generated and implemented. This sends the message that continuous improvement is important: that it is a way of life in the Lean organization.

Use team-based measures; avoid measures of individual performance. Though this likely goes against every management bone in your body, you need to realize that individual measures can conflict with the collaborative, team nature of the Lean organization. Remember the concept of pull, which we covered in Chapter 4, and the possible reallocation of work to different people based on demand. If people are measured solely on individual performance, this may create a conflict with the pull system in practice. Similarly, if you have implemented office cells, as described in Chapter 3, the message of any implemented performance measures must be that it is a single-process, single-team concept. Issues with individual performance can be identified and addressed through other means, such as observation during a gemba walk.

Whatever performance measures are selected, performance information must be available in a timely fashion. People will disregard performance measures that are not timely, and rightfully so, as they tend to be less actionable. Therefore, obtaining and summarizing performance data must be simple. For example, it may be more practical to track the number and nature of quality problems as they arise than to calculate a percentage of "defects per opportunity." Remember, an aspect of Lean thinking is to identify nonstandard conditions. This can apply to the subject of performance measurements as well.

The timeliness of reviewing performance leads to a discussion of "management timeframe" or "pitch." This is defined as the frequency at which the system will be reviewed to ensure that performance expectations are being met. Lean organizations have very short management timeframes. Performance is reviewed frequently. In a Lean organization, you need to address performance issues very quickly. There is one thing that cannot be recouped in life, and that is time. Throughout this book, I have shown you various visual methods of monitoring performance. In the call center example, the management timeframe was near-real time. In other situations, it was hourly, daily, or even weekly. Do not confuse frequent performance assessment as a form of "micromanagement," most will recognize it for its true intent—to identify problems in a timely manner so that you can correct them quickly, avoiding the loss of people's valuable and limited time.

Recognition

Remember, recognition was the second most important factor in work satisfaction in Herzberg's list. Leaders need to take advantage of opportunities to provide recognition as warranted. It can be as simple as a "thank you" for a job well done or a more formal celebratory event. Usually, some type of celebration will cap off any major improvement effort, such as a kaizen event. During these celebrations, you can recognize team members for their contributions but, again, simpler expressions of recognition can be very effective over time. For example, if you show appreciation to someone who identified a problem, you are reinforcing the message that identifying problems is a good thing in the Lean organization. Another simple method might be to post comments of recognition on the visual management boards, perhaps on those displaying performance measures or continuous improvement efforts.

When you recognize appropriate behaviors, you reinforce the desired characteristics of a Lean culture. Although it is true that organizations consist of many people, it is the leader or leaders who define culture through words and actions on a daily basis.

Summary

As a leader in a Lean organization, your role is vastly different from the historic management role in traditional organizations. It might take some time for you and other managers to transition to this new role. *And* it requires the support of your own manager. Leading a Lean organization takes perseverance, consistency, and discipline, but you will find the work environment to be much more rewarding. Gone will be the days of directing the most basic tasks, replaced with a much more rewarding role of developing the most important asset of any organization, its people.

The Quality Toolbox

When you are implementing Lean—or *any* kind of continuous improvement initiative—you need a full toolbox. The late Kaoru Ishikawa, a professor of engineering at Tokyo University, identified seven basic quality tools—visual tools focused on data interpretation and analysis, key components of implementing Lean. Over the years, this list has been modified, but the tools most commonly identified as the basics are

- Flowcharts
- Run charts
- Histograms
- Pareto charts
- Cause-and-effect diagrams
- Control plans
- Control charts

These basic quality tools can be very helpful when implementing Lean in an office and service environment and to continuous improvement in general. You need to be familiar with them and recognize when opportunities to apply them arise.

Flowcharts

A flowchart is a graphic display of the step-by-step flow of information, service, or product. Each step is displayed in the sequence in which it occurs. Symbols can be used to highlight the nature of a particular step. For example, a diamond-shaped symbol is most often used to display decision-making steps. A "D"-shaped symbol can highlight delays in the process.

Process data can be added to the map to increase its value in terms of the amount of information that it conveys.

Flowcharts can take several different formats. The most common is displayed in Figure A.1.

Another common flowchart format is called a *spaghetti diagram* (see Figure A.2). The spaghetti diagram displays the physical flow of the information, service, or product. A layout of a facility or an area within a facility serves as a backdrop for the map, and the location at which each step is performed is noted on the map. Arrows then display the flow. This type of flowchart can highlight issues arising from the current location at which steps are performed, which may not be obvious in the flowchart format shown in Figure A.1.

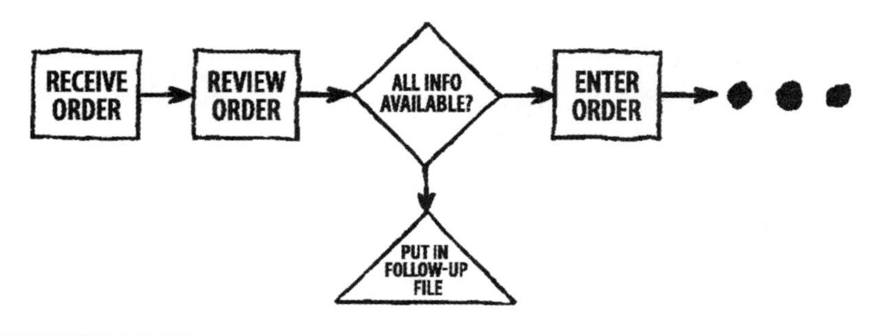

Figure A.1 Traditional flowchart.

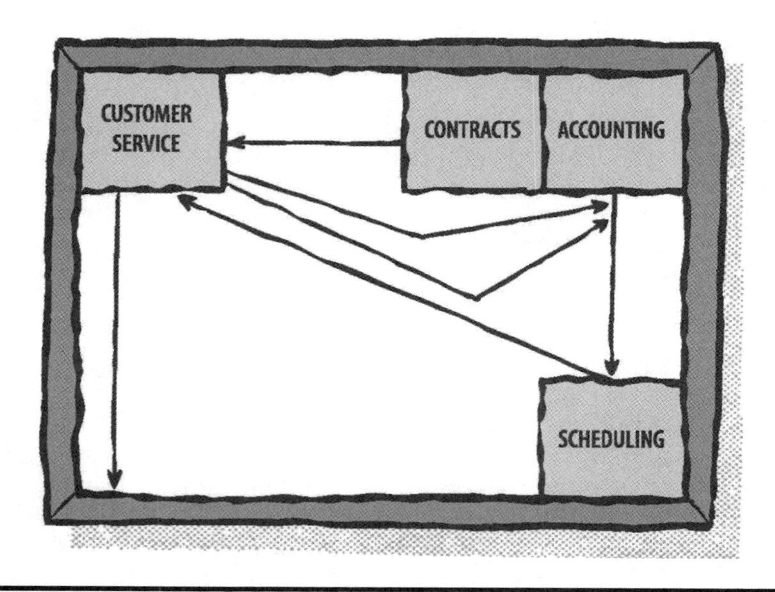

Figure A.2 Spaghetti diagram in an office or service context.

Run Charts

A run chart displays data in the sequence in which it occurs *over time* (see Figure A.3). Almost any measurable data can be displayed in this way, which is why they are used so often. Run charts can illustrate even subtle changes in data over time. Therefore, it is imperative to display the data in the sequence in which it actually occurred to avoid potential misinterpretations. In the example in Figure A.3, patient wait time increased throughout the day from the first patient treated to the 50th. The question that this facility should be asking is "Why?"

Histograms

A histogram is a form of frequency chart created by counting (orders, for example). The same numerical data used to create a run chart can be used to create a histogram (see Figure A.4). Although trends over time can be seen on a run chart, you cannot readily see such changes on a histogram. However, you can identify other issues with histograms. In the example provided, the number of orders that were shipped in a given timeframe (number of days) is counted. For example, five orders were shipped in 10 to 11 days. Ten orders were shipped in 12 to 13 days, and so on. The chart shows that

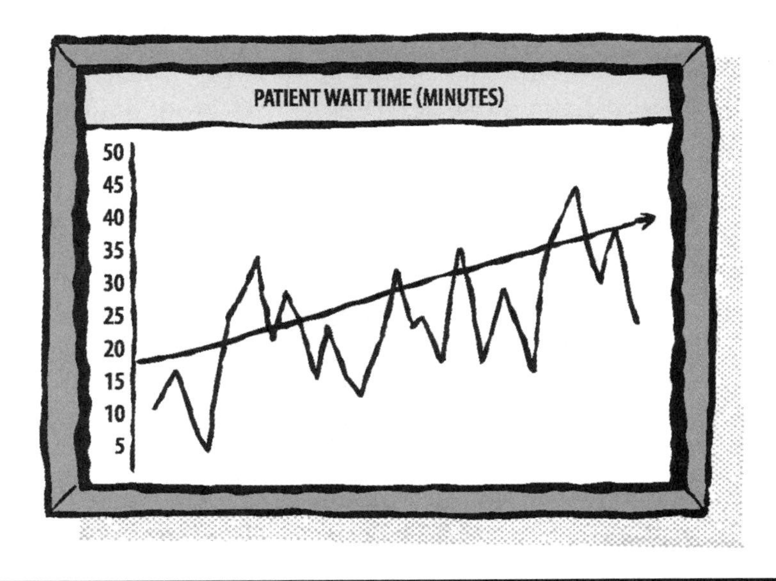

Figure A.3 Run chart example.

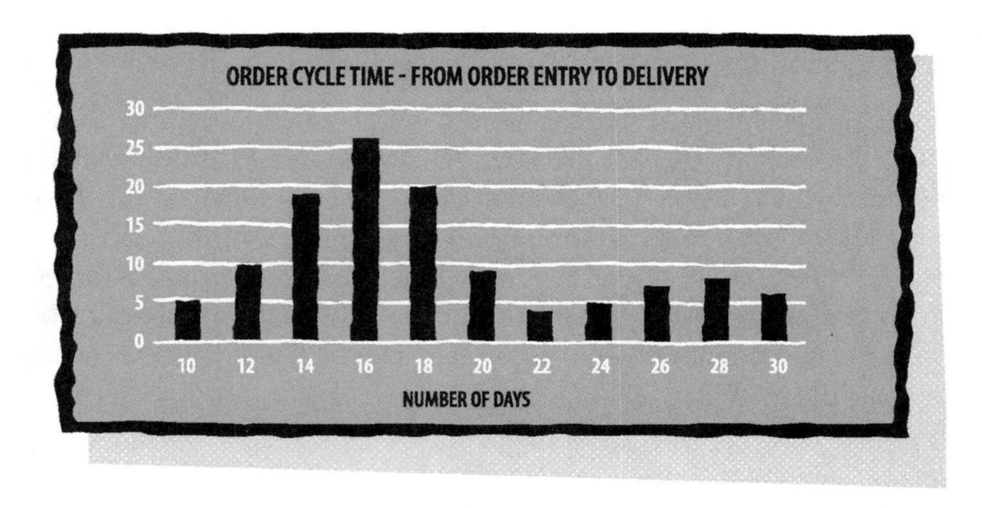

Figure A.4 Example of a histogram.

most orders ship between 10 and 22 days, with the majority shipping at the 16- to 17-day mark. However, the chart shows that there is a group of orders that ship in 24 to 30 days. Once this data is made visible with the histogram, the organization needs to ask itself what makes those orders different.

Pareto Charts

A Pareto chart is another form of frequency chart, also created by counting. However, in Pareto charts, the information is presented in such a way as to highlight severity, from most frequent to least frequent occurrence (see Figure A.5).

Cause-and-Effect Diagram

Cause-and-effect diagrams visually display all of the possible causes for a problem. The problem is the "effect." Once you identify the effect, you can begin to identify various possible causes, often using the 5 Whys technique. There are two common formats for cause and effect diagrams, both shown in Figure A.6. The one on the left is called a "fishbone," or Ishikawa, diagram. The one on the right is a "block diagram." Both convey the same information, but in a different format. Your personal preference will dictate which format you use.

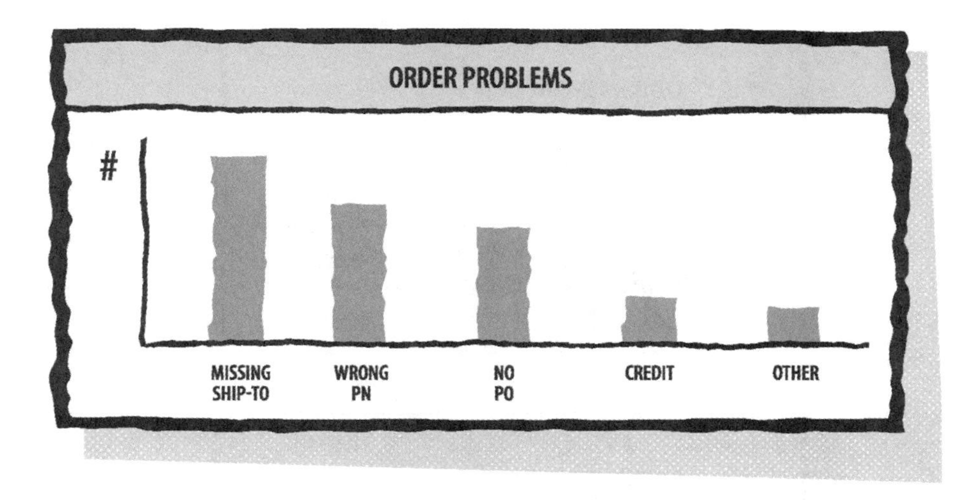

Figure A.5 Example of a Pareto chart.

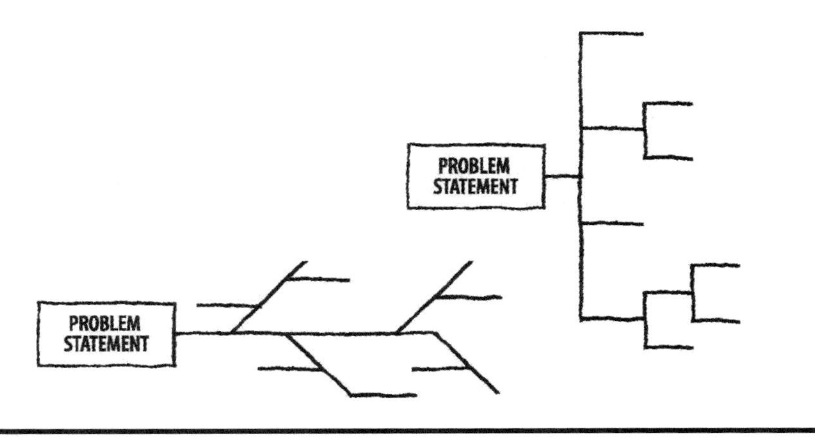

Figure A.6 Examples of cause-and-effect diagrams.

Control Plans

Control plans represent the steps that need to be followed to ensure that a process performs to expectations over time. Standard work instructions are part of a control plan. However, other steps may also be taken to ensure that a process remains in control over time. For example, audits of compliance to standard work can be part of a control plan. Periodic measurement of the process is usually part of a control plan, as well. Preventive techniques should be part of control plans. For example, an organization might implement a document control system that ensures that only the most recent versions of documents are being used. And finally, periodic "refresher"

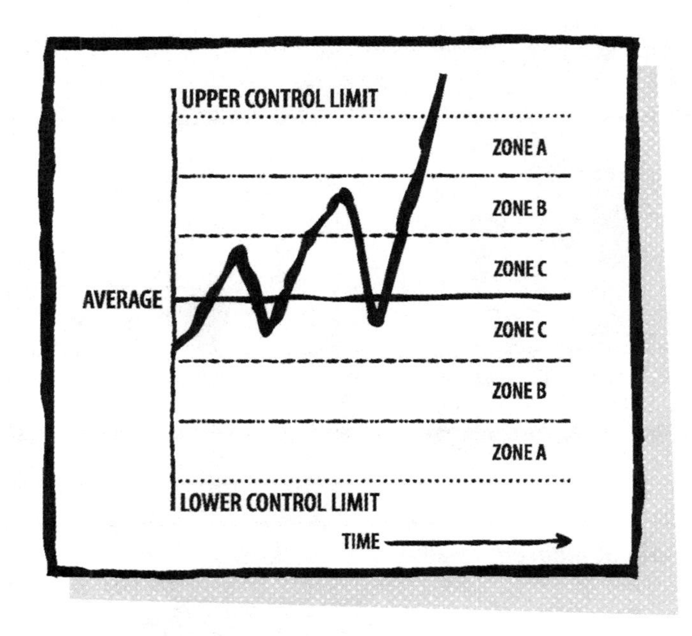

Figure A.7 Example of control chart.

training can be part of a control plan. Taken together, these elements constitute a plan that can help ensure that a process performs as expected over time.

Control Charts

Control charts provide a visual means of monitoring a process over time and to identify when action should be taken to return a process to a more desirable level of performance. Control charts are essentially run charts with additional information displayed. "Control limits" are displayed on the run charts as horizontal lines. Control limits define how the process is expected to perform. Based on the current process performance and where it falls with respect to these limits (the various zones on the chart), the chart can tell people when to take action on the process or when to leave it alone (see Figure A.7). Control charts are particularly helpful for processes in which even subtle changes in the process can have a substantial negative impact on performance or the customer. If deemed helpful, they can be used as part of the previously mentioned control plan.

Forms

The following forms have proven useful in many Lean office and service implementation.

Job Breakdown Sheet

Operation: ______________________________

Tools and Materials Needed: ______________________________

Common Key Points: ______________________________

General Quality Standards: ______________________________

#	**Important Steps (What)** A logical segment of the operation in which something happens to advance the work, including expected time to complete	**Key Points (How)** - Quality (will create a defect) - Safety (can cause injury) - Efficiency (special information to make work easier)	**Reasons (Why)** Reasons for the key points

Workplace Scan Checklist

Number of Problems	Rating Level	Date	Date	Date	Date	Date
5 or more	Level 0					
3 to 4	Level 1					
2	Level 2					
1	Level 3					
None	Level 4					

Category	Item	Level				
Sort	**Distinguish between what is needed and not needed**					
	Unneeded equipment, tools, furniture, etc., are present					
	Unneeded items are on walls, bulletin boards, etc.					
	Items are present in aisle ways, stairways, corners, etc.					
	Unneeded inventory, supplies, documents, or materials are present					
	Safety hazards (electrical wires, boxes, machines) exist					
Set in Order	**A place for everything and everything in its place**					
	Correct places for items are not obvious					
	Items are not in their correct places					
	Aisle ways, workplaces, equipment locations are not indicated					
	Items are not put away immediately after use					
	Height and quantity limits are not set (in storage areas)					
Shine	**Cleaning, and looking for ways to keep clean and organized**					
	Floors, walls, stairs, and surfaces are not free of dirt, etc.					
	Equipment is not kept clean and free of dirt, etc.					
	Cleaning materials are not easily accessible					
	Labels, signs, etc., are not clean and unbroken					
	Other cleaning problems (of any kind) are present					
Standardize	**Maintain and monitor the first three categories**					
	Necessary information is not visible or available					
	Checklists do not exist for all cleaning and maintenance jobs					
	All quantities and limits are not easily recognizable					
	File retention rules are not being followed					
	How many items cannot be located in 30 seconds?					
Sustain	**Stick to the rules**					
	How many workers have not had 5S training					
	How many times are personal belongings not easily stored?					
	How many times last week was daily 5S not performed?					
	How many out-of-date procedures, forms, etc., can be found?					
	How many times last quarter were monthly 5S inspections not performed?					
	Total					

Workplace Scan Display

Target Area: ________________	
Stakeholders: ________________	
Activities: ________________	
Area Map and Arrow Diagram	**Scan Checklist**
Before/After Photos	**Actions List**

Office Waste Walk Form

Waste Category	Observation	Suggested Action	Impact on Cost, Service, Quality, or Safety
Overproduction Processing more information than is needed, sooner than is needed by the next process. Examples: filled "in-boxes," excessively detailed information, printing information before needed.			
Inventory Processing information in excess of "one-piece flow." Examples: batch processing, periodically performing a task, computer programs run over night.			
Correction Inaccurate or missing information and the "inspection" of work. Examples: rework, returning information to a previous step for correction, requests for clarification, errors found later in the process.			
Extra Processing Unnecessary or overly time consuming processes or steps. Examples: excessive approvals, steps not required by the "customer," process time variation between workers.			

Office Waste Walk Form

Waste Category	Observation	Suggested Action	Impact on Cost, Service, Quality, or Safety
Waiting Information in queue and not being worked on. Examples: filled "in-boxes"—paper or electronic, information waiting on decisions, approvals, resources to work on it, etc.			
Motion Unnecessary or excessive movement of people. Examples: walking to and from faxes, central filing, printers, supply cabinets, meetings, disorganized work areas, searching for information, etc.			
Transportation Unnecessary or excessive movement of information—manual or electronic. Examples: multiple handoffs, e-mail, inter-office mail, unclear process flow, excessive "cc's," poor office layout, etc.			
Underutilized People Not using people's full skills and abilities—mental, creative, and physical. Examples: limited job responsibilities causing excessive hand-offs, multiple approvals, no "back-up" resources, people not assisting each other, etc.			

Physical Layout
(Current or Proposed)

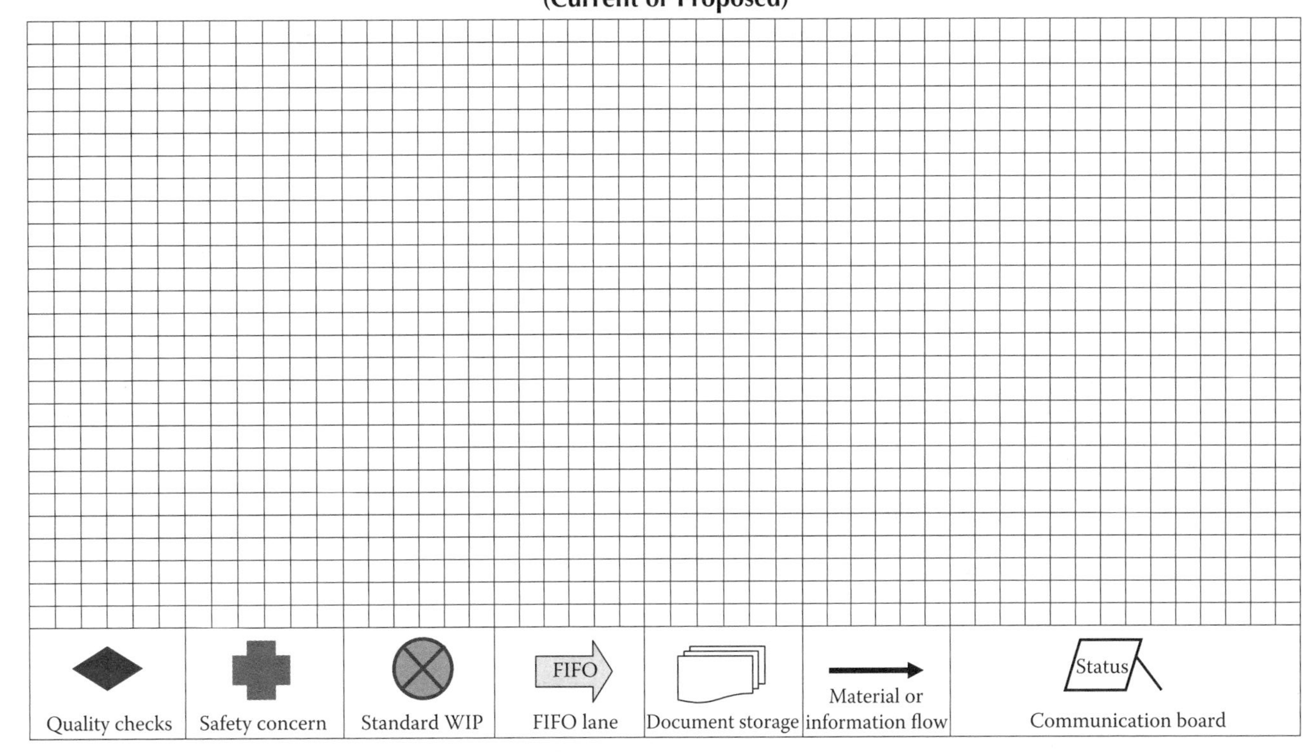

Directions:
(1) Using the grid above, locate each work station and/or piece of equipment (e.g., machines, faxes) in the current or proposed layout. Also, locate any storage areas (e.g., materials, file cabinets).

(2) Draw arrows depicting the material and/or information flow between work stations. Calculate the current or expected distance travel.

(3) Note locations of special activities and conditions such as quality checks, safety concerns, standard work-in-process (e.g., FIFO lanes), and positions of visual communication boards.

Time Observation Form

Date: | Event or observer:

Process: | Takt time:

#	Component task	1	2	3	4	5	6	7	8	9	10	11	12	13	14	15	16	17	Component task time	Special comments and points observed
	Total cycle time																			

Operator Balance Chart

Time

Operations

Directions:

(1) List each operation across the bottom of the chart (x-axis).

(2) Determine the appropriate time scale along the left side (y-axis). Locate the Takt Time on the y-axis and draw a line across the chart.

(3) Draw a "bar" representing the process time for each operation.

(4) Compare the process time for each operation to the Takt Time. Identify operations that are greater than Takt Time and less than Takt Time.

Standard Work Combination Sheet

Date:	Work sequence:				Demand:	Manual ——— Auto - - - - - - Walking
Part name:					Takt Time:	
Step #	Work content description	Time			Work Content Graph (sec/div.)	
		Manual	Auto	Walk		

Work Station Requirements Form

Process: ____________________

Work Station	Activities	Equipment/Software/Tools/Supplies	Information

Directions:

(1) For each Work Station, identify the activities that are expected to be performed there.

(2) List all of the resources that will be required to perform the activities. Be certain to include all equipment, tools, materials, and information (instructions, standards). Consider how and where these resources will be stored at the Work Station. Remember that proper location of all required resources can significantly reduce non-valued-added time.

Cause and Effect Diagram

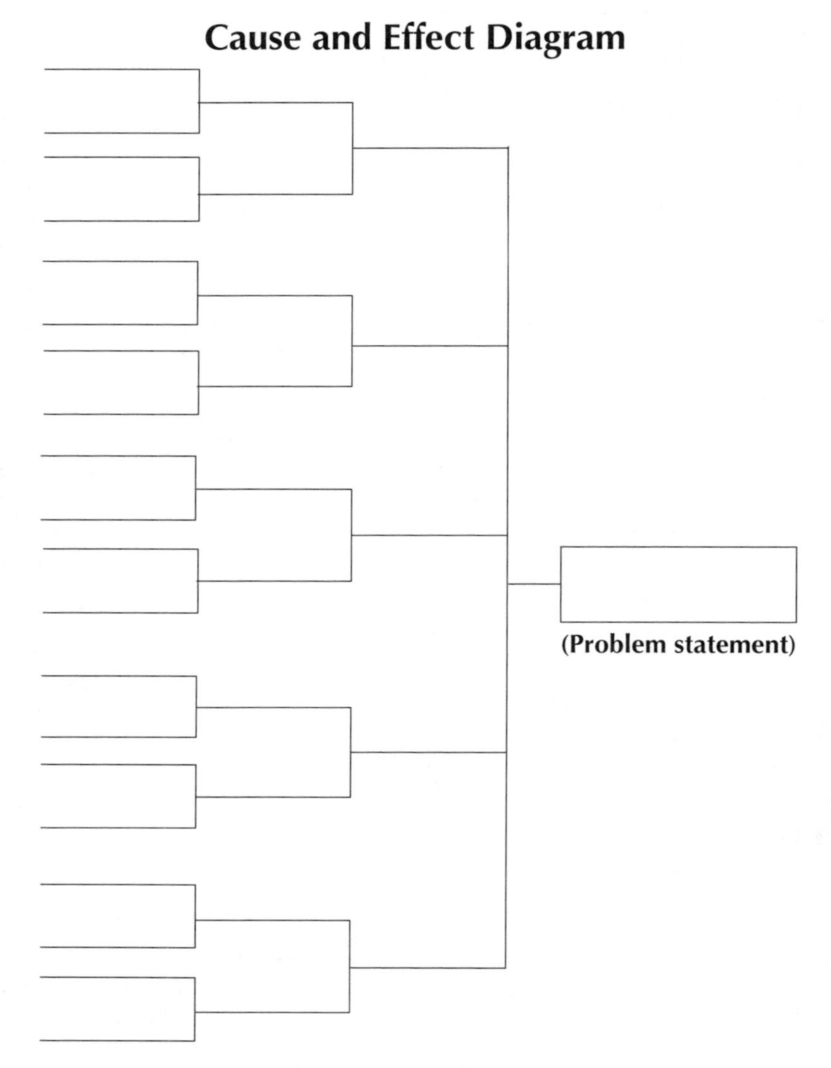

Directions:

(1) Using the figure above as a guide, create a C&E Diagram on a flip chart, dry erase board or some other surface. On the right hand side, write a clear statement of the problem. Use available data if available. Avoid unclear statements.

(2) Identify any appropriate "categories" of causes for each major "branch" on the diagram. Examples of categories are People, Process, Policies, Systems, Environment, Materials, and Measures, or others that seem appropriate for the problem to be addressed.

(3) Working from branch to branch, identify potential causes for the problem. Be certain to ask the "5 Why's" to identify root causes. Add additional "levels" of "branches" on the diagram as necessary.

(4) It may be desirable to "weight" the various root causes by estimated impact on creating the stated problem.

Mistake Proofing Chart

<table>
<tr><td colspan="6">Problem Statement or Defect:</td></tr>
<tr><td colspan="6">Root Causes or Errors:</td></tr>
<tr><td rowspan="5">Root Causes:
(5 Why's)</td><td></td><td></td><td></td><td></td><td></td></tr>
<tr><td></td><td></td><td></td><td></td><td></td></tr>
<tr><td></td><td></td><td></td><td></td><td></td></tr>
<tr><td></td><td></td><td></td><td></td><td></td></tr>
<tr><td></td><td></td><td></td><td></td><td></td></tr>
<tr><td colspan="6">Devices:</td></tr>
<tr><td rowspan="3">Improvement Ideas:
Level 3: Detect Defect
Level 2: Detect Error
Level 1: Eliminate Error</td><td></td><td></td><td></td><td></td><td></td></tr>
<tr><td></td><td></td><td></td><td></td><td></td></tr>
<tr><td></td><td></td><td></td><td></td><td></td></tr>
<tr><td colspan="6">Types of Mistake Proofing Devices:
1) Guide/Reference/Interference Pin
2) Template
3) Limit switch/micro-switch
4) Counter
5) Odd-part-out-method
6) Sequence restriction
7) Standardize and solve
8) Critical condition indicator
9) Detect delivery chute
10) Stopper/gate
11) Sensor
12) Mistake proof your mistake proofing device</td></tr>
</table>

Directions:

(1) Using the figure above as a guide, create a Mistake Proofing Chart on a flip chart, dry erase board, or some other surface. At the top, write a clear statement of the problem or defect. Use available data if available. Avoid unclear statements. Be sure to distinguish between a "defect" and an "error." The defect is the result.

(2) Identify potential causes or errors for the problem or defect. Be sure to ask the 5 Why's to identify root causes. Transfer this information over from a Cause and Effect Diagram, if one was created.

(3) For each root cause, identify improvement ideas—specific "devices" that can be implemented to address the cause or error. For each, identify Level 1 devices wherever possible. Avoid solely relying on Level 3 or even Level 2 devices.

Index

A

Accounting, 109–116
 function improvement, 115–116
 Generally Accepted Accounting Practices, 110
 management board, 115
 stability issues, 110
 standardizing of processes, 110–114
 visual, 114–115
Achievement, job satisfaction and, 131
Acknowledgments, sending, 33, 35
Activities, value stream organization, 1–8
 cross-functional teams, 3–4
 departmental roles, defining by value stream, 4–6
 functional approach, cell approach, contrasted, 4
 plan for every process, example of, 7
Activity identification, flow system design, 30
Actual place. *See* Gemba
Advancement, job satisfaction and, 131
Art department board, 56, 58
A3 management process, 137

B

Batch and queue processing
 vs. combining activities, 23
 vs. cross-functional, colocated teams, 25
Belonging, job satisfaction and, 130
Benefit of pull systems, 58–60
Boss, relationship with, 131

C

Cause-and-effect diagram, 147–147, 160
 examples of, 147
Cell approach, functional approach, contrasted, 4
Changeover speed, 93–96
Characteristics of pull systems, 45
Clarity, job satisfaction and, 136
Color file system, 80
Combining activities, 23–25
Company policy, 131
Completion of tasks, visual indicators, 70
Comprehensive visual project management system, 67
Concurrent performance, parallel performance of activities, 29
Concurrent processing, 28–29
Contact method, mistake proofing, 87
Continuous flow processing, with multiple roles, 25–28
Continuous improvement in visual management, 72–73
Control chart, 148
 example of, 148
Control plans, 147–148
Counter devices, mistake proofing, 89
Creation of standard work, steps, 17–19
Credibility, job satisfaction and, 136
Critical condition indicator devices, mistake proofing, 91
Critical thinking, 136
 job satisfaction and, 136

Critical Thinking: Its Definition and Assessment, 136
Cross-functional teams, 3–4
Cross-training needs, flow system design, 36
Customer service, lean application, 116–122
function improvement, 120–122
process standardization, 118–119
stability issues, 116–118
visual, 119–120

D

Decision rules
establishing, 48–50
posted by queue, 49
Defining rules for queue, 45, 58
Defining standard work, 10–11
Delays, waste from, 19
Delivery detection devices, mistake proofing, 91–92
Demand rate
determination, flow system design, 30–32
inclusion in standard work, 14
Departmental roles, defining by value stream, 4–6
Design of flow systems, 29–39
activity identification, 30
cross-training needs, 36
demand rate determination, 30–32
resource requirements determination, 32–34
responsibilities, identification of, 34–36
roles, identification of, 34–36
training, 36
visual management techniques, 37
Detection method categories, mistake proofing, 87
Devices, mistake proofing, 88–93
Dissatisfaction with job, factors, 131
Driving continuous improvement, 131–134

E

Elements of standard work, 12–14
Environment of workplace, employee satisfaction with, 38
Equation for takt time, 30–32
Escalation process, problem board with, 73
Esteem, impact of, 130
External activities, internal activities, distinguishing, 94
Extra processing, waste from, 19

F

Fairness, job satisfaction and, 136
Fisher, Alec, 136
5S's, 75–96
color file system, 80
contact method, mistake proofing, 87
counter devices, mistake proofing, 89
critical condition indicator devices, mistake proofing, 91
delivery detection devices, mistake proofing, 91–92
detection method categories, mistake proofing, 87
devices, mistake proofing, 88–93
fixed-value method, mistake proofing, 87
guide/reference/interference device, mistake proofing, 88
internal activities, external activities, distinguishing, 94
light contact electrical devices, mistake proofing, 88–89
mistake proofing, 84–93
motion-step method, mistake proofing, 87
odd-part-out devices, mistake proofing, 89–90
origin of, 77
quick changeover, 93–96
sensor devices, mistake proofing, 93
sequence restriction devices, mistake proofing, 90
setup reduction, 93–96
standardize/solve devices, mistake proofing, 90–91
stopper/gate, mistake proofing, 92–93
streamlining results, 95
supermarket pull system, office supplies, 81
template/checklist devices, mistake proofing, 88

visible shine standard, 82
workplace organization, 76–84
Five whys, 132
Fixed-value method, mistake proofing, 87
Flow in office, service, 21–39
acknowledgments, sending, 33, 35
approaches, 22–29
batch and queue processing
vs. combining activities, 23
vs. cross-functional, colocated teams, 25
changes, sustained periods of, 32
combining activities, 23–25
concurrent performance, parallel performance of activities, 29
concurrent processing, 28–29
continuous flow processing, with multiple roles, 25–28
flow systems design, 29–39
activity identification, 30
cross-training needs, 36
demand rate determination, 30–32
resource requirements determination, 32–34
responsibilities, identification of, 34–36
roles, identification of, 34–36
training, 36
visual management techniques, 37
invoices, 33, 35
lead time reductions, 38
office cell, example of, 27
order design, 33, 35
order entering, 33, 35
process balance, 26
process time reductions, 38
quality improvements, 38
resource needs determination, 35
standard work combination form, 36
takt image, 37
takt time, equation for, 30–32
value stream mapping, 24
workplace environment, employee satisfaction with, 38
Flowchart, 143–144
traditional, 144
Forms, 149–161
Forms of pull systems, 42–45
characteristics, 45
sequential pull systems, 42
supermarket pull systems, 42
Functional approach, cell approach, contrasted, 4

G

GAAP. *See* Generally Accepted Accounting Practices
Gemba, 138–139
Gemba walks, 138
Generally Accepted Accounting Practices, 110
Growth, job satisfaction and, 131
Guide/reference/interference device, mistake proofing, 88

H

Herzberg, Frederick, 130–131
Hierarchy of needs, 130
Histogram, 145–146
example of, 146
HR. *See* Human resources
Human resources, lean application, 122–128
function improvement, 126–128
stability issues, 123–124
standardizing processes, 124–125
visual, 125–126
visual management display, 126

I

Identification of means, visibility, pull systems, 55–56
Identification of roles, responsibilities, flow system design, 34–36
Implementation of pull systems, 54–58
limits for queue, establishing, 56–57
location identification, 55
monitoring, 57–58
rules for queue, defining, 58
training pull system procedures, 57
visibility, identification of means, 55–56

Individuals, activities for, value stream organization, 1–8
cross-functional teams, 3–4
departmental roles, defining by value stream, 4–6
functional approach, cell approach, contrasted, 4
plan for every process, example of, 7
Instruction, steps in, 135
Internal activities, external activities, distinguishing, 94
Inventory waste, 18
Invoices, 33, 35

J

Job breakdown sheet, 150
Job enrichment characteristics, 131
Job instruction, 135
Job methods, 134

K

Kanbans, 45, 50, 54
Kepes, György, 61
Kettering, Charles, 132

L

The Language of Vision, 61
Lead time reductions, 38
Leadership, 129–142
A3 management process, 137
accuracy, 136
achievement, 131
advancement, 131
belonging, 130
boss, relationship with, 131
breadth, 136
clarity, 136
company policy, 131
credibility, 136
critical thinking, 136
depth, 136
dissatisfaction with job, factors, 131
driving continuous improvement, 131–134
esteem, 130
fairness, 136
Fisher, Alec, 136
five whys, 132
gemba, 138–139
gemba walks, 138
growth, 131
Herzberg, Frederick, 130–131
hierarchy of needs, 130
instruction, steps in, 135
job enrichment characteristics, 131
job instruction, 135
job methods, 134
Kettering, Charles, 132
logic, 136
love, 130
Maslow, Abraham, 130–131
mentoring, 134–137
non-value-adding activities, identification of, 138
peers, relationship with, 131
performance measurement, 139–141
physiological needs, 130
Plan-Do-Check-Act improvement cycle, 132
precision, 136
recognition, 131, 141
relevance, 136
responsibility, 131
safety, 130
salary, 130–131
satisfaction with job, factors, 131
Scriven, Michael, 136
self-actualization, 130
Shook, John, 137
significance, 136
supervision, 131
work conditions, 131
work itself, 131
Lean applications, 97–128
accounting, 109–116
function improvement, 115–116
Generally Accepted Accounting Practices, 110
stability issues, 110
standardizing of processes, 110–114
visual, 114–115
visual management board, 115

customer service, 116–122
function improvement, 120–122
process standardization, 118–119
stability issues, 116–118
visual, 119–120
human resources, 122–128
function improvement, 126–128
stability issues, 123–124
standardizing processes, 124–125
visual, 125–126
visual management display, 126
marketing, 98–105
function improvement, 103–105
inside sales visual management, 104
process standardization, 100–102
stability issues, 99–100
tailoring standard work, 101
visual, 102–103
purchasing, 105–109
function improvement, 109
stability issues, 106–107
standardizing, 107–108
visual, 108–109
visual management techniques, 108
sales, 98–105
function improvement, 103–105
inside sales visual management, 104
process standardization, 100–102
stability issues, 99–100
tailoring standard work, 101
visual, 102–103
Lean Lexicon, 1, 42
Lean Thinking, 21
Lean tools, 75–96
color file system, 80
contact method, mistake proofing, 87
counter devices, mistake proofing, 89
critical condition indicator devices, mistake proofing, 91
delivery detection devices, mistake proofing, 91–92
detection method categories, mistake proofing, 87
devices, mistake proofing, 88–93
fixed-value method, mistake proofing, 87
guide/reference/interference device, mistake proofing, 88
internal activities, external activities, distinguishing, 94
light contact electrical devices, mistake proofing, 88–89
mistake proofing, 84–93
motion-step method, mistake proofing, 87
odd-part-out devices, mistake proofing, 89–90
origin of, 77
quick changeover, 93–96
sensor devices, mistake proofing, 93
sequence restriction devices, mistake proofing, 90
setup reduction, 93–96
standardize/solve devices, mistake proofing, 90–91
stopper/gate, mistake proofing, 92–93
streamlining results, 95
supermarket pull system, office supplies, 81
template/checklist devices, mistake proofing, 88
visible shine standard, 82
workplace organization, 76–84
Level pull in office, 41–60
art department boards, 56, 58
benefit of pull systems, 58–60
decision rules for queue, establishing, 48–50
decision rules posted by queue, 49
defined rules for queue, 45
forms of pull systems, 42–45
characteristics, 45
sequential pull systems, 42
supermarket pull systems, 42
kanbans, 45, 50, 54
leveling output in increments, 53
leveling system, 52–54
light as kanban in call center, 50
limits defined for queue, 45
limits on queues, establishing, 46–48
overproduction, 42
pull system implementation, 54–58
limits for queue, establishing, 56–57
location identification, 55
monitoring, 57–58

rules for queue, defining, 58
training pull system procedures, 57
visibility, identification of means, 55–56
sequential pull systems, 45
single point of management, 59
timed baskets in pull system, 51
value stream, implementing pull system through, 47
value stream mapping icons, pull systems using, 43
visibility of queues, 45–46
worker managed flag as pull signal, 52
worker managed visual signals, 45, 50–52
Leveling output in increments, 53
Leveling system, 52–54
Light as kanban in call center, 50
Light contact electrical devices, mistake proofing, 88–89
Limits, queues, establishing, 46–48, 56–57
Limits defined for queue, 45
Location identification, pull systems, 55
Logic, job satisfaction and, 136

M

Management, 129–142
activities performed in area, 65–66
approaches, 63–64
completion of tasks, visual indicators, 70
comprehensive visual project management system, 67
continuous improvement in visual management, 72–73
elements of, 64–72
escalation process, problem board with, 73
evaluation, 69–71
factors impacting, 63
function of area, 65
methodology, 68–69
performance expectations, 71–72
performance measurement board, 70
"plan for every process," with visual indicators, 67
project scheduling board, 67
purpose of area, 65
Supplier-Input-Process-Output-Customer, 65–66
example of, 66
task identification, 66–68
traffic light techniques, project status board using, 71
visual, 61–74
voice of customer, 66
Managing to Learn, 137
Marketing, lean application, 98–105
function improvement, 103–105
inside sales visual management, 104
process standardization, 100–102
stability issues, 99–100
tailoring standard work, 101
visual, 102–103
Maslow, Abraham, 130–131
Means identification, visibility, pull systems, 55–56
Mentoring, 134–137
Mistake proofing, 84–93
contact method, 87
counter devices, 89
critical condition indicator devices, 91
delivery detection devices, 91–92
detection method categories, 87
devices, 88–93
fixed-value method, 87
guide/reference/interference device, 88
light contact electrical devices, 88–89
motion-step method, 87
odd-part-out devices, 89–90
sensor devices, 93
sequence restriction devices, 90
standardize/solve devices, 90–91
stopper/gate, 92–93
template/checklist devices, 88
Mistake proofing chart, 161
Monitoring pull systems, 57–58
Motion-step method, mistake proofing, 87
The Motivation to Work, 130
Movement of office/service personnel, waste from, 19
Multitasking environment, standard work in, 15

N

Needs hierarchy, 130
Non-value-added activities, 18–19
Non-value-adding activities, identification of, 138
NVA activities. *See* Non-value-adding activities

O

Odd-part-out devices, mistake proofing, 89–90
Office cell, example of, 27
Office flow, 21–39
  acknowledgments, sending, 33, 35
  approaches, 22–29
  batch and queue processing
    vs. combining activities, 23
    vs. cross-functional, colocated teams, 25
  changes, sustained periods of, 32
  combining activities, 23–25
  concurrent performance, parallel performance of activities, 29
  concurrent processing, 28–29
  continuous flow processing, with multiple roles, 25–28
  flow systems design, 29–39
    activity identification, 30
    cross-training needs, 36
    demand rate determination, 30–32
    resource requirements determination, 32–34
    responsibilities, identification of, 34–36
    roles, identification of, 34–36
    training, 36
    visual management techniques, 37
  invoices, 33, 35
  lead time reductions, 38
  office cell, example of, 27
  order design, 33, 35
  order entering, 33, 35
  process balance, 26
  process time reductions, 38
  quality improvements, 38
  resource needs determination, 35
  standard work combination form, 36
  takt image, 37
  takt time, equation for, 30–32
  value stream mapping, 24
  workplace environment, employee satisfaction with, 38
Office movement, waste from, 19
Office pull system, 41–60
  art department boards, 56, 58
  benefit of pull systems, 58–60
  decision rules for queue, establishing, 48–50
  decision rules posted by queue, 49
  defined rules for queue, 45
  forms of pull systems, 42–45
    characteristics, 45
    sequential pull systems, 42
    supermarket pull systems, 42
  kanbans, 45, 50, 54
  leveling output in increments, 53
  leveling system, 52–54
  light as kanban in call center, 50
  limits defined for queue, 45
  limits on queues, establishing, 46–48
  overproduction, 42
  pull system implementation, 54–58
    limits for queue, establishing, 56–57
    location identification, 55
    monitoring, 57–58
    rules for queue, defining, 58
    training pull system procedures, 57
    visibility, identification of means, 55–56
  sequential pull systems, 45
  single point of management, 59
  timed baskets in pull system, 51
  value stream, implementing pull system through, 47
  value stream mapping icons, pull systems using, 43
  visibility of queue of work, 45
  visibility of queues, 45–46
  worker managed flag as pull signal, 52
  worker managed visual signals, 45, 50–52
Office/service, standard work, 9–20
  benefits, 16–17
  correction waste, 18

definition of standard work, 10–11
demand rate inclusion in, 14
elements of, 12–14
extra processing waste, 19
inventory waste, 18
job instruction, 16
methodology, 12–13
multitasking environment, example of, 15
non-value-added activities, 18–19
overproduction waste, 18
personnel movement, waste from, 19
purpose of, 10–11
rationale, 12–13
service personnel movement, waste from, 19
standard work instruction, example of, 14
standardizing creative work, 20
steps to creating, 17–19
time, 13–14
timing, 13–14
transportation waste, 19
underutilized persons, waste from, 19
virtually displaying, 14–19
waiting, waste from, 19
waste activities, 18–19
Office waste walk form, 153–154
Operator balance chart, 157
Order design, 33, 35
Order entering, 33, 35
Organization by value stream, 1–8
cross-functional teams, 3–4
departmental roles, defining by value stream, 4–6
functional approach, cell approach, contrasted, 4
plan for every process, example of, 7
Overproduction, 18, 42

P

Pareto chart, 146
example of, 147
PDCA improvement cycle. *See* Plan-Do-Check-Act improvement cycle
Peers, relationship with, 131
Performance expectations, 71–72
Performance measurement, 139–141
Performance measurement board, 70
Physical layout (current or proposed), 155
Physiological needs, 130
Plan-Do-Check-Act improvement cycle, 132
Plan for every process
example of, 7
with visual indicators, 67
Practical Lean Accounting, 114
Precision, job satisfaction and, 136
Procedures for pull system, training in, 57
Process balance, 26
Process time reductions, 38
Project scheduling board, 67
Pull system implementation, 41–60. *See also* Office pull system
limits for queue, establishing, 56–57
location identification, 55
monitoring, 57–58
rules for queue, defining, 58
training pull system procedures, 57
visibility, identification of means, 55–56
Purchasing, lean application, 105–109
function improvement, 109
stability issues, 106–107
standardizing, 107–108
visual, 108–109
visual management techniques, 108

Q

Quality improvements, 38
Quality toolbox, 143–148
cause-and-effect diagram, 147–147
examples of, 147
control chart, 148
example of, 148
control plans, 147–148
flowchart, 143–144
traditional, 144
histogram, 145–146
example of, 146
Pareto chart, 146
example of, 147
run chart, 145
example, 145
spaghetti diagram, 144

R

Recognition, job satisfaction and, 131, 141
Resource needs determination, 35
Resource requirements determination, flow system design, 32–34
Responsibility, job satisfaction and, 131
Roles
  departmental, defining by value stream, 4–6
  identification of, flow system design, 34–36
Rules for queue, defining, 58
Run chart, 145
  example, 145

S

Safety issues, 130
Salary, job satisfaction and, 130–131
Sales, lean application, 98–105
  function improvement, 103–105
  inside sales visual management, 104
  process standardization, 100–102
  stability issues, 99–100
  tailoring standard work, 101
  visual, 102–103
Satisfaction with job, factors in, 131
Satisfaction with workplace environment, 38
Scriven, Michael, 136
Self-actualization, 130
Sensor devices, mistake proofing, 93
Sequence restriction devices, mistake proofing, 90
Sequential pull systems, 42, 45
Service, standard work, 9–20
  benefits, 16–17
  correction waste, 18
  defining standard work, 10–11
  demand rate inclusion in, 14
  elements of, 12–14
  extra processing waste, 19
  instruction, example of, 14
  inventory waste, 18
  job instruction, 16
  methodology, 12–13
  in multitasking environment, example of, 15
  non-value-added activities, 18–19
  office/service personnel movement, waste from, 19
  overproduction waste, 18
  purpose of, 10–11
  rationale, 12–13
  service personnel movement, waste from, 19
  standardizing creative work, 20
  steps to creating, 17–19
  time, 13–14
  timing, 13–14
  transportation waste, 19
  underutilized persons, waste from, 19
  virtually displaying, 14–19
  waiting, waste from, 19
  waste activities, 18–19
Service flow, 21–39
  acknowledgments, sending, 33, 35
  approaches, 22–29
  batch and queue processing
    vs. combining activities, 23
    vs. cross-functional, colocated teams, 25
  changes, sustained periods of, 32
  combining activities, 23–25
  concurrent performance, parallel performance of activities, 29
  concurrent processing, 28–29
  continuous flow processing, with multiple roles, 25–28
  flow systems design, 29–39
    activity identification, 30
    cross-training needs, 36
    demand rate determination, 30–32
    resource requirements determination, 32–34
    responsibilities, identification of, 34–36
    roles, identification of, 34–36
    training, 36
    visual management techniques, 37
  invoices, 33, 35
  lead time reductions, 38
  office cell, example of, 27

order design, 33, 35
order entering, 33, 35
process balance, 26
process time reductions, 38
quality improvements, 38
resource needs determination, 35
standard work combination form, 36
takt image, 37
takt time, equation for, 30–32
value stream mapping, 24
workplace environment, employee satisfaction with, 38
Service personnel movement, waste from, 19
Setup reduction, 93–96
Shingo, Shigeo, 86
Shook, John, 137
Single point of management, 59
SIPOC. *See* Supplier-Input-Process-Output-Customer
Skills matrix, example, 127
Spaghetti diagram, 144
Standard work combination form, 36
Standard work combination sheet, 158
Standard work for office/service, 9–20
benefits, 16–17
correction waste, 18
defined, 10–11
demand rate inclusion in, 14
elements of, 12–14
extra processing waste, 19
instruction, example of, 14
inventory waste, 18
job instruction, 16
methodology, 12–13
multitasking environment, example of, 15
non-value-added activities, 18–19
office movement, waste from, 19
overproduction waste, 18
purpose of, 10–11
rationale, 12–13
standardizing creative work, 20
steps to creating, 17–19
time, 13–14
timing, 13–14
transportation waste, 19
underutilized persons, waste from, 19
virtually displaying, 14–19
waiting, waste from, 19
waste activities, 18–19
Standardize/solve devices, mistake proofing, 90–91
Standardizing creative work, 20
Steps to creating standard work, 17–19
Stopper/gate, mistake proofing, 92–93
Streamlining results, 95
Supermarket pull systems, 42, 81
Supervision, 131
Supplier-Input-Process-Output-Customer, 65–66
example of, 66
Sustained periods of change, 32

T

Takt image, 37
Takt time, equation for, 30–32
Task identification, 66–68
Teams, cross-functional, 3–4
Template/checklist devices, mistake proofing, 88
Time observation form, 156
Timed baskets in pull system, 51
Traffic light techniques, project status board using, 71
Training, flow system design, 36
Training pull system procedures, 57
Transportation waste, 19

U

Underutilized persons, waste from, 19

V

Value stream, 6–7
implementing pull system through, 47
organization by, 1–8
cross-functional teams, 3–4
departmental roles, defining by value stream, 4–6
functional approach, cell approach, contrasted, 4
plan for every process, example of, 7

Value stream mapping, 24
  icons, 43
Virtually display, standard work, 14–19
Visibility, identification of means, pull systems, 55–56
Visibility of queues, 45–46
Visible shine standard, 82
Visual management, 61–74
  activities performed in area, 65–66
  approaches, 63–64
  completion of tasks, visual indicators, 70
  continuous improvement in, 72–73
  elements of, 64–72
  escalation process, problem board with, 73
  evaluation, 69–71
  factors impacting, 63
  function of area, 65
  management system, 67
  methodology, 68–69
  performance expectations, 71–72
  performance measurement board, 70
  "plan for every process," with visual indicators, 67
  project scheduling board, 67
  purpose of area, 65
  Supplier-Input-Process-Output-Customer, 65–66
    example of, 66
  task identification, 66–68
  traffic light techniques, project status board using, 71
  voice of customer, 66
Visual management techniques, flow system design, 37
VOC. *See* Voice of customer
Voice of customer, 66

W

Walks, gemba, 138
Waste, activities causing, 18–19
Work conditions, 131
Work itself, job satisfaction and, 131
Work station requirement form, 159
Worker managed flag as pull signal, 52
Worker managed visual signals, 45, 50–52
Workplace environment, employee satisfaction with, 38
Workplace scan checklist, 151
Workplace scan display, 152

Z

Zero Quality Control: Source Inspection and the Poka-Yoke System, 86

About the Author

Drew Locher is currently managing director for Change Management Associates (CMA). CMA provides various business improvement consulting and organizational development services to industrial and service organizations. CMA will assist organizations to successfully implement systems and quality management principles within their operations to improve business performance.

Since 1986, Drew has been working to implement innovative Business Management strategies in a wide spectrum of business environments. From 1986 to 1990 he worked to develop and deliver Business Improvement programs for General Electric. During this time, he had the opportunity to gain firsthand experience in significant business improvement initiatives throughout a large industrial and service company. In 1990, Drew left GE to form CMA. CMA is a consortium of individuals who share a similar business management and improvement vision. They are committed to seeing innovative strategies implemented in all business environments. The CMA approach is practical and application oriented - a 'learn as you're doing and succeeding' approach.

In 1997, CMA partnered with the National Institute of Standards and Technologies', Manufacturing Extension Partnership, to develop a "Lean University". For more information, check out the NIST/MEP Lean Website. Since 2001, Drew has proudly been a faculty member of the Lean Enterprise Institute (LEI), the not-for-profit organization of James P. Womack, the co-author of the landmark book Lean Thinking,. LEI are committed to educating organizations worldwide in the concepts of Lean through its publications and workshops. He is also an adjunct faculty member at Ohio State University delivering Lean related workshops as part of the Fisher School of Business.

In 2004, Drew co-authored a book titled, The Complete Lean Enterprise – Value Stream Mapping for Office and Administrative Processes (Productivity

Press). In April 2005, the book won the prestigious Shingo Prize for Excellence in Manufacturing. In 2008, he authored Value Stream Mapping for Lean Development: A How-to Guide to Streamline Time to Market (Productivity Press).

Drew received a Bachelor of Science degree from the University of Delaware in Mechanical Engineering, as well as a Master of Science degree from Drexel University in Electrical and Computer Engineering. He has also received a Master of Business Administration from Cornell University.